# 幻獣の製作方法

## 可動式幻獣玩偶の
## 製作技法＆作品集

### 綺想造形蒐集室

## 【幻獸】

這是指擁有神秘或奇妙力量的虛構生物。

## 【可動式玩偶】

這是指運用金屬線等材料構成骨骼，讓手腳、脖子和尾巴等部位可以活動的動物類玩偶。

# Contents

〔製作注意事項〕使用有機溶劑、塗料和接著劑時，請務必保持環境通風。作業會產生粉塵時，請穿戴防塵面罩。

# The Winter Fox   Anya Boz

「Room Guardians」是藝術家Anya Boz最廣為人知的藝術玩偶。
「Room Guardians」是一群可隨意活動擺出姿勢的動物，牠們以
水晶之心守護在房間的你，阻擋來自惡靈的侵害。
Anya Boz在2011年開始製作Room Guardians，並且在2014年轉為
專業的藝術家，至今已創作出數百隻的動物作品。其中一隻
「The Persian Cat Room Guardian（波斯貓房間守護者）」，為
2016年極受歡迎的網路迷因（※）素材。
※網路迷因是一種在圖片加上標題的模板素材圖，由歐美地區帶起的一股熱潮。

Anya Boz現居紐約布魯克林區，在自家兼工作室中從事創作，並
且透過網路銷售作品。她不曾就讀專業學校，而是花了數年的時
間，透過自學不斷嘗試與探究，形成一套製作方法。她製作的手
法並非基於完美的設計圖，而是任由創作自行發展，所以直到開
始動手製作，都無法預知最終成品將會是甚麼模樣。

Anya Boz is an artist that is best known for her art dolls called Room Guardians.
Room Guardians are poseable animals that have a crystal heart to protect your room
from evil spirits.
She created her first room guardian in 2011 and became a full time artist in 2014.
She has since gone on to create hundreds of room guardians over the years, one of
which (The Persian Cat Room Guardian) became a beloved internet meme in 2016.
She currently lives in Brooklyn, New York where she creates her room guardians in
her home studio and sells most of her work online. Anya was never classically trained
in school. All her methods were developed over years of personal experimentation
and exploration. You will not see any patterns or designs before she begins creating
because she prefers to work without a fully developed plan to let the room guardian
create itself as she goes.

[HP] https://www.anyabozartist.com

[Instagram] @anyaboz

[YouTube]
https://www.youtube.com/channel/UCsEmd9Y92XZiQtYNEaOtj9w

# The Winter Fox的製作方法

## 骨架製作

### 工具和材料

①直徑約3mm的娃娃骨架
　（約40cm）
②直徑約1.5mm的娃娃骨架
　（約18cm）
③線徑0.9mm的鍍鋅細金屬
　網（網眼大小約2.5cm）
④鋪棉
　（約4cm寬、約120cm）

⑤織線（約360cm）
⑥老虎鉗
⑦小石頭
⑧喜歡的水晶（小顆）
⑨線徑約1.2mm的銅線
⑩熱塑性塑膠（約180ml）
⑪鋁箔紙

① Approximately 16 inches of ⅛ inch ball and socket doll armature
② Approximately 7 inches of 2/32 inch ball and socket doll armature
③ A square of 20 gauge 1 inch mesh galvanized chicken wire
④ Approximately 4 feet of quilt batting cut into rough 1.5 inch strip

⑤ Approximately 12 feet of yarn any color
⑥ Needle nose pliers
⑦ One small stone
⑧ One small crystal of your choice
⑨ 18 gauge copper wire
⑩ Approximately 6 oz of Instamorph plastic pellets
⑪ A roll of aluminum foil

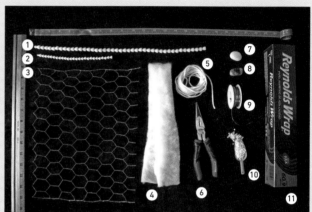

## Memo

可動式布偶骨架（娃娃骨架）
這是一種用來製作布偶的骨架，由如蛇彎曲的球狀塑膠關節組合而成。這種骨架和金屬線相比，不但多次彎折都不會折斷，還很容易維持形狀姿態，所以很推薦大家使用。脖子或較粗的尾巴可使用直徑約3mm的規格，腳或較細的尾巴可使用直徑約1.5mm的規格。由於無法彎折成90度，所以不會用於手臂。金屬線的可動範圍較廣，所以比較適合用於小型手臂。

Plastic Doll Armatures
These armatures are made specifically for doll making. They consist of plastic interlocking ball joints that bend like a snake. They are generally better than using wire because they do not wear out over time and they hold position much better. I use ⅛ inch for most room guardian necks and thick tails. I use 3/32 inch for legs and thin tails. I don't like to use them for arms because they don't bend at a 90 degree angle. Wire has a better range of motion for small arms.

## 製作步驟

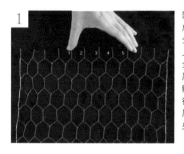

製作身體的細金屬網是由金屬線纏繞而成。這次為了方便示範說明，將纏繞的部分塗成紅色，只有一根金屬線的部分則塗上黃色和藍色。細金屬網的上部是纏繞的金屬線（紅色金屬線）。基本上將紅色金屬線呈縱向垂直的方向，組合成玩偶的身體（因為這樣比水平方向的架構穩定）。從細金屬網的上部中央，找出5個紅色金屬線網當成中線軸心，不完全位於正中央也沒關係。

Shaping the chicken wire to form the body
Chicken wire is made up of wires that are twisted together. I color coated my chicken wire so it's easier to see. The twisted parts are painted red, and the single wires are painted yellow and blue depending on which way they angle. You want the top of your chicken wire square to be a set of twisted wires (red wires). You do not have to use my method exactly because it can be hard to keep track of, but always build your room guardian with red twisted wires running vertically along the body, not horizontal. It will be much more stable that way.
In the center of your chicken wire, count out 5 red prongs. It's okay if it is not exactly in the center.

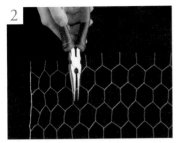

在步驟 1 確定了中線軸心的紅色金屬線區塊後，用老虎鉗剪斷區塊左側的藍色金屬線。

Using needle nose pliers, cut the top of the blue wire on the left of your chosen 5 red prongs.

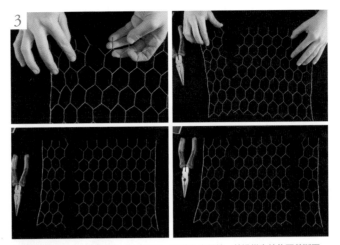

在步驟 2 剪斷藍色金屬線後，也剪斷下一條藍色金屬線。就這樣直線往下剪斷下一條黃色和藍色的金屬線。接著剪斷步驟 1 中央紅色金屬線區塊右側的黃色金屬線。同樣直線往下剪斷黃色和藍色的金屬線。

Next, cut the blue wire opposite the wire you just cut. Now cut down in a straight line, alternating from yellow to blue as you go. Next cut the yellow wire on the right side of your five red prongs. And cut in a straight line alternating blue and yellow.

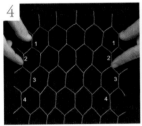 

將細金屬網彎折捲成筒狀。使兩側的紅色金屬線重疊。

Your chicken wire must be curled into a cylinder. The corresponding red wires should meet when you fold the sides together.

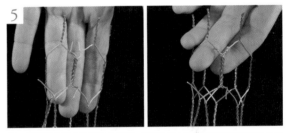

將一邊剪斷未固定的黃色金屬線，捲繞在兩邊都固定未剪斷的黃色金屬線。藍色金屬線也以相同方法捲繞固定，使細金屬網相連固定，並且用老虎鉗輔助繞緊。為了避免受傷，將金屬線的尖端（切口）往筒狀內側捲入。

To join them, wrap the loose yellow wires around the corresponding yellow wires, and the loose blue wires around the corresponding blue wires. Tighten them with your pliers. Make sure to wrap the sharp end facing inside so you avoid scratching your fingers.

不斷重複步驟 5，直到紅色金屬線的 4 個地方都相互結合。

Repeat this until you have joined four red wires together.

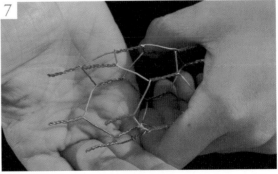

筒狀金屬線骨架完成。步驟 1 中設定為中線軸心的紅色金屬線，為筒狀的上部。這個紅色金屬線部分就是安裝脖子骨架的位置。

Now that we have a cylinder, we can wrap up the 5 red prongs on top. This is where we will eventually attach the neck armature.

**8**

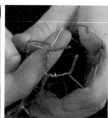

彎折第1根紅色金屬線，使其與相鄰的第2根紅色金屬線相接的藍色金屬線結合。用老虎鉗將第1根紅色金屬線彎折固定。

Take one red prong and bend it over the blue wire attached to the second red prong. Fold it closed with your pliers.

**9**

彎折第2根紅色金屬線，使其與相鄰的第3根紅色金屬線相接的藍色金屬線結合。重複步驟8～9，直到剩下最後一根紅色金屬線。

Now take that second red prong and wrap it over the blue wire attached to the third red prong. Repeat this until you have one loose red prong.

**10**

為了讓娃娃骨架穿過，最後一根紅色金屬線不捲繞固定，只先彎折。

This one, just bend down without wrapping. We will need it to open up later when we fit the ball and socket armature into the wires.

**11**
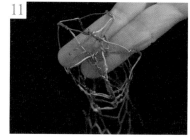

金屬線如照片般呈星形形狀。

The wires should form a star shape.

**12**
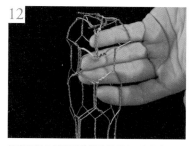

找出紅色金屬線重疊捲繞的部分，這個部分是玩偶的胸部中心。將重疊的紅色金屬線往下壓，使周圍藍色和黃色的金屬線呈V字形。

Next, find the double red wire we wrapped earlier. This marks where the middle of the chest on our room guardian will be. Press the double red wire down until the blue and yellow wires form a V shape.

**13**

剩下的紅色金屬線如照片般彎折。星形部分呈現往內凹的形狀。

Next bend the rest of the red wires so the star shape is concave.

**14**
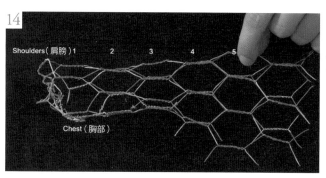

以步驟12構成胸部的金屬線為基準，找出背後肩膀往右的第5根紅色金屬線。

Using the chest wire as a guide, find the middle red wire five sections down from the shoulders along the back.

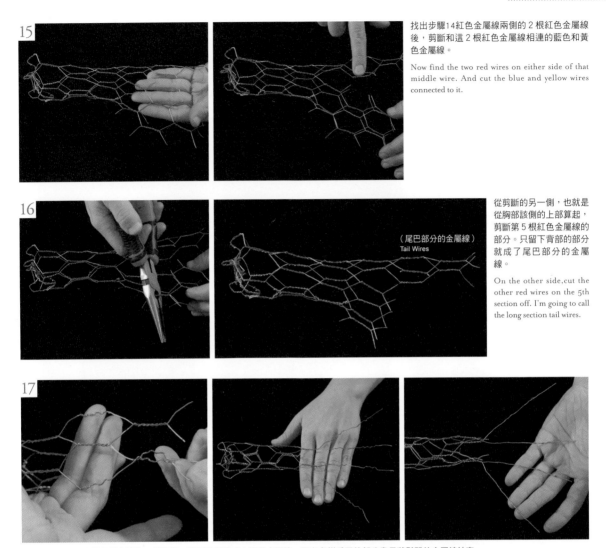

找出步驟14紅色金屬線兩側的2根紅色金屬線後，剪斷和這2根紅色金屬線相連的藍色和黃色金屬線。

Now find the two red wires on either side of that middle wire. And cut the blue and yellow wires connected to it.

（尾巴部分的金屬線）
Tail Wires

從剪斷的另一側，也就是從胸部該側的上部算起，剪斷第5根紅色金屬線的部分。只留下背部的部分就成了尾巴部分的金屬線。

On the other side, cut the other red wires on the 5th section off. I'm going to call the long section tail wires.

尾巴部分的金屬線鬆開最右邊約2個網眼。鬆開後就變成4條長金屬線。用老虎鉗扁平的部分盡量將鬆開的金屬線拉直。

Take the tail wires and unwind the bottom two sections. There should be four long wires. Straighten these wires as much as possible using the flat part of the pliers.

相對於尾巴部分的另一側金屬線，分別為4根單邊鬆開的藍色和黃色金屬線。

There are now four loose blue and yellow wires.

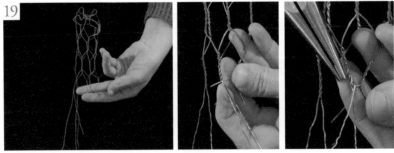

將最左邊的黃色金屬線末端穿過左側2根尾巴的金屬線之間，捲繞在紅色金屬線上。

Take the leftmost yellow prong. And poke it between the left two tail wires. And wrap it around the red wire.

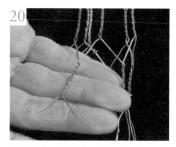

**20**

相反的另一側藍色金屬線也依照步驟19的
方式處理。

Do the same thing with the blue wire on the
opposite side.

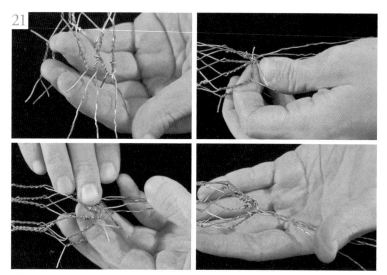

**21**

重複步驟19～20，直到沒有藍色和黃色的金屬線需要穿過尾巴金屬線。

Then keep wrapping the loose blue and yellow wires around the red wires until there are no more wires
poking out.

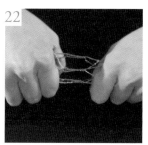

**22**

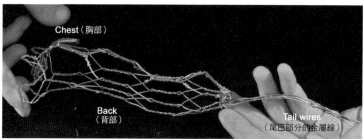

Chest（胸部）

Back
（背部）

Tail wires
（尾巴部分的金屬線）

如此一來就可以將金屬線塑造成自己喜歡的形狀。這次想將背部做成S曲線的造型。在之後的成形作業中，會包裹上鋁箔紙，並且黏
貼上布料。這些步驟會增加相當多分量，所以最好盡量將金屬線骨架做成細窄的體型。至此細金屬網的塑造作業完成。

VNow you can take your hands and shape it how you want. I want mine to have an S shape curve on the back. It's best to keep this stage as thin as
possible. Foil wrapping and fur can add a lot of bulk. The chicken wire stage is finished.

**23**

將直徑約3mm的娃娃骨架剪成自己喜歡的長度。將照片中短的骨架用
於脖子，長的骨架用於尾巴。照片中的娃娃骨架有11個連接部件，但
是創作者後來想做長一點的脖子，而又增加了2個連接部件。將步驟
10中，尚未捲繞固定，只有彎折的紅色金屬線拉起，將脖子用的娃娃
骨架穿入星形的正中央。

Next split the ⅛ armature into a short piece for the neck and a long piece for
the tail. You can make these as long as you want. This neck is 11 joints long, but
I ended up adding two more joints to it for a longer neck. Lift up the one loose
red wire you left at the top of the chicken wire body and push the neck armature
into the middle of your star pattern.

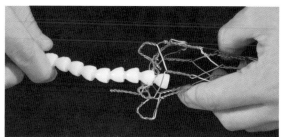

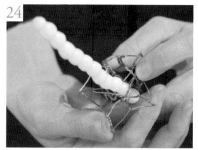

**24**

將金屬線往下壓，用
周圍的金屬線固定娃
娃骨架的1個連接部
件。

Press the wire back
down, locking one joint
of the armature inside
the wires.

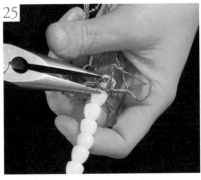 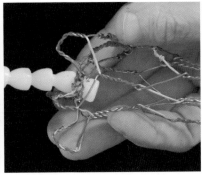

接著用別的金屬線纏繞，用老虎鉗盡量將金屬線繞緊。牢牢固定住娃娃骨架避免晃動。

Wrap it around the other wires and clamp it closed with your pliers. Try to get the neck armature as snug as possible. You might need to tighten some of the other wires so it doesn't wiggle around.

將直徑約1.5mm的娃娃骨架剪成兩段相同的長度。這些將當成腳使用。

Split your 3/32 armature in two equal pieces. These will be the legs.

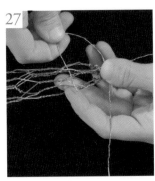

拉出2根尾巴金屬線，沿著背部金屬線彎折。將腳固定在背部金屬線骨架。安裝在背部，腳的可動範圍會比安裝在身體前面廣。

Pull Two of your tail wires out. And wind it into one section of the armature along the middle of the back. We will attach the legs to the back of the armature because it will give them a better range of movement than attaching them to the front.

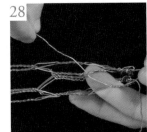 

用尾巴金屬線牢牢捲繞住1個腳部娃娃骨架的連接部件。

Coil that around the first joint of one leg armature until it is snug.

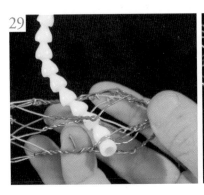 

將剩餘的尾巴金屬線捲繞在細金屬網骨架。另一隻腳也用相同方法捲繞固定。

And then wrap the excess wire back into the chicken wire armature. Repeat on the next leg.

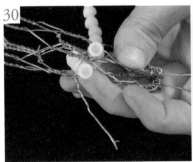 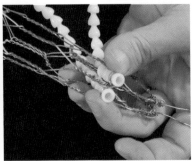

若只用尾巴金屬線無法牢牢固定，可以用剩餘的細金屬網或其他種類的金屬線固定。

If using the tail wires doesn't work out for you, you can use some of the chicken wire scraps or any type of wire to keep them in place.

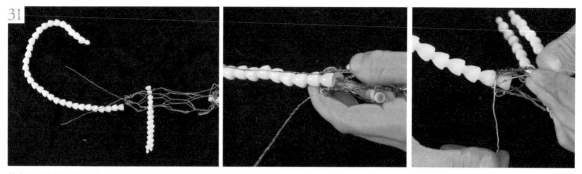

將步驟23的尾巴娃娃骨架，插入尾巴金屬線的根部，並且用剩餘的2根尾巴金屬線緊緊捲繞第1個連接部件。

Now take the rest of the ⅛ armature for the tail. Insert it at the base of the tail wires, and wrap the remaining two tail wires around the first joint until it is snug.

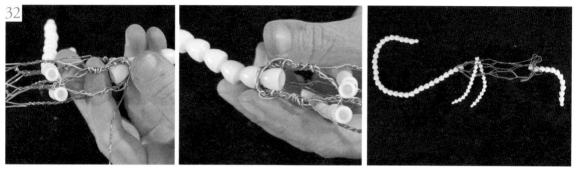

將剩餘的金屬線捲繞在細金屬網骨架。

Then wrap the excess into the chicken wire armature.

將熱塑性塑膠（商品名稱：Instamorph）放入容器中，淋上熱水直到塑膠變成透明色。用老虎鉗或夾子，將軟化的熱塑性塑膠從熱水夾出。經過數分鐘冷卻後，用手揉成球狀。

Next, pour your Instamorph pellets into a bowl, and pour boiling water over them until they turn clear. Using your pliers or tongs, take the soft Instamorph out of the water. Let it cool for a few seconds and then knead it into a ball in your hands.

將熱塑性塑膠撕下一半後，從下往上按壓黏接在脖子娃娃骨架。也可以將熱塑性塑膠再次放回熱水中，使其更加軟化後，再進行這個步驟。只要完全包覆住第1個連接部件和金屬線的星形部分即可，不需要包裹得很漂亮。

Pull it in half, and press one half over the neck armature pushing from the bottom up. For the best results, put the plastic back in the hot water before this step so it's extra soft. It doesn't have to look pretty. Just make sure the plastic covers the first ball joint and wraps fully around some of the wires in the star shape.

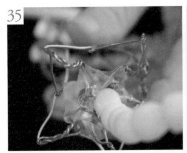
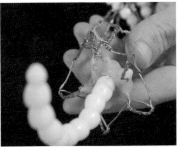

**35**

可看出星形金屬線部分完全固定在熱塑性塑膠中。熱塑性塑膠變白硬化後，請確認娃娃骨架脖子的周圍連接部分是否會鬆動。若連接部分不牢固，可以再次將熱塑性塑膠部分放入熱水中，重新塑形。

You can see how the star is fully embedded in the plastic and not just stuck on top of it. When the plastic hardens and turns white, test the neck armature to make sure it does not feel loose within the plastic. If you feel it is not stable enough, just pour hot water over the plastic again to reshape it.

**36**

將步驟34的熱塑性塑膠再撕成一半，用和脖子相同的方式，固定尾巴娃娃骨架。

Next, break the remaining instamorph in your bowl in half, and I repeat the process on the tail using one part of the instamorph.

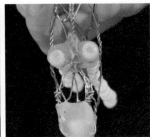
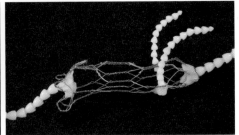

**37**

用步驟36剩下的熱塑性塑膠固定腳。相較於其他部位，用金屬線纏繞腳的部分較少，所以較不易穩固，不過腳要支撐身體，所以必須牢牢接合。另外，直到熱塑性塑膠硬化前，很難確定部件是否完全接合固定。因此若為了確認腳的接合狀況，而在塑膠完全硬化前觸碰部件，就可能使熱塑性塑膠部分鬆動。因此一定要在塑膠完全硬化變白後，才確認腳的固定狀況。

Use the other half of the Instamorph for the legs. The legs are a little more difficult because they have less wire for support, but they need to be very stable to hold up the body. It can be hard to tell if they are fully stable until the Instamorph is solid. Resist the urge to test the legs before it is fully hardened because that can loosen the plastic's hold on them. Wait until the plastic is white, then test for any wiggle before remelting again.

**38**

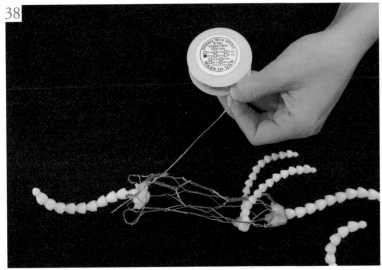

參考骨骼結構，先測量手臂需要的長度。銅線多餘的部分可以纏繞在細金屬網骨架上，所以可先將銅線裁切得比實際所需長度長一些。這次所需的銅線長度大約為15cm。一條裁成所需長度，另一條裁成所需長度的2倍後先對折。

Using the armature as a reference, measure how long you want the arm to be. Always cut the wire a little bit longer than you want because we need excess wire to wrap into the chicken wire. These wires measured 6 inches. Cut one wire the length you need, then double one wire and fold it in half.

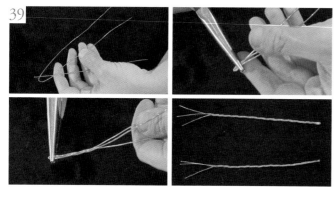

**39**

將 1 條銅線穿過對折銅線的對折處後，用老虎鉗將對折銅線纏繞固定住 1 條銅線。用老虎鉗壓對折處，用另一隻手將銅線捲繞在一起。為了纏繞在骨架上，末端先不要捲繞在一起。將銅線平均鬆鬆地纏繞在一起。即便纏繞成束，纏繞時也不要過於施力，使銅線維持有點鬆的程度。

Feed the single wire into the loop of the double wire. Then clamp the loop around the single wire with your pliers. Get a tight hold on the clamped loop with your pliers and begin twisting the wires together with your other hand. Leave the ends untwisted so you can wrap them into the armature later. When twisting your wires, you want to twist them evenly and loosely. They should be just tight enough to stay together but loose enough that you are not putting stress on them.

# 手部塑造

## 工具和材料

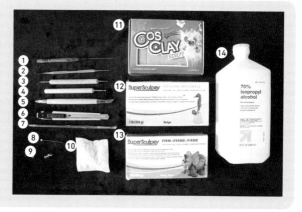

①細節針
②塑形刀（蠟刀）
③球狀黏土刮刀（小）
④球狀黏土刮刀（大）
⑤矽膠型黏土刮刀
⑥美工刀
⑦畫筆（舊的）

①Needle Clay Tool
②Wax Carver Spatula Tool
③Small Ball Stylus
④Large Ball Stylus
⑤Silicone Tipped Sculpting Tool
⑥Utility Knife
⑦Old Paint Brush

⑧珠針
⑨直徑 6～8 mm 玻璃珠
⑩砂紙（180號）
⑪彈性軟陶黏土
⑫軟陶黏土（米色／一般）
⑬軟陶黏土（灰色／硬質）
⑭消毒酒精

⑧Quilting Pin
⑨6 to 8mm Glass Beads
⑩180 grit Sandpaper
⑪Cosclay
⑫Super Sculpey Original
⑬Super Sculpey Firm
⑭Rubbing Alcohol

## 製作步驟

**1**

只用彈性軟陶黏土（商品名稱：Cosclay）製作手臂。若只使用彈性軟陶黏土難以塑形，也可以混合軟陶黏土（商品名稱：Super Sculpey）。彈性軟陶黏土（商品名稱：Cosclay）較乾燥，容易碎裂，所以要揉捏至塊狀成團。

For the hands, I am going to sculpt with pure Cosclay. Feel free to mix some Super Sculpey into your hands if that makes sculpting easier for you. Cosclay can be very crumbly fresh out of the box. But a little bit of rolling and kneading should get it to a good consistency.

**2**

用力揉捏成小的筒狀後，將一端捏成手套狀。利用手指調整形狀，做出大概的手掌、拇指和手腕的形狀。用珠針劃出刻痕，做出手指形狀。先在中央劃出左右對稱的刻痕。

Once your Cosclay is kneaded, roll out a small cylinder, and form a mitten shape on one side. Refine the shape with your fingers until you have a rough palm, thumb, and wrist formed. Take your quilting pin and cut out the fingers. Start in the center to keep them even.

**3**

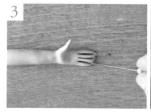

接著再劃出每根手指的刻痕，並且用自己的手指，輕輕搓揉分割後的部分。太長的手指，則捏去一點黏土再次塑形。

Take each finger and roll it gently in your fingers to round out the edges. If some of the fingers are too long at this point, just pinch off the tips and smooth them down again.

**4**

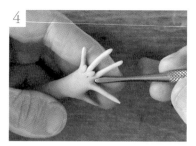

將極小的球狀黏土添加在指縫間，用塑形刀接合按壓出指蹼的形狀。

Add a tiny ball of clay between each finger. Smooth it out to create the finger webbing with the small paddle of the wax carver.

**5**

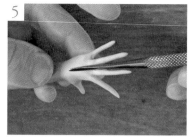

按壓指間，做出關節形狀。

Press down between the fingers to create knuckles.

**6**

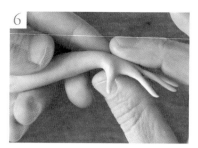

輕輕彎折手腕，做出想要的姿勢。

Gently pose the wrist the way you want it.

**7**

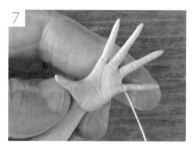

用珠針劃出手指紋路和掌紋。

Draw the lines in the hand using your quilting pin.

**8**

為了讓手掌以及關節平滑，塗上薄薄的消毒酒精。請注意，不要將畫筆浸泡在酒精太久。

Smooth the palm and the knuckles with your rubbing alcohol. Coat the hand lightly. Avoid over soaking the brush.

**9**

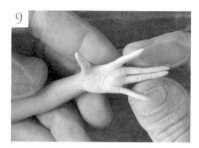

整體表面變得平滑後，做出手部姿勢。中指和無名指一起擺出相同姿勢，會更靈動。

Once your hand is thoroughly smoothed out, you can begin posing it. You can press the middle finger and ring finger together to pose them as one to create dramatic poses.

**10**

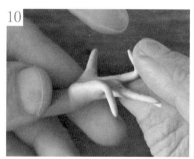

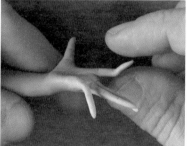

固定姿勢前，建議先將所有手指彎成直角。實際的手指會有兩處彎折，但這裡簡單化處理，只在手指正中央一處彎折。即便姿勢是完全伸直的手指，都先折出折痕再伸直。這樣一來，即便直線都會呈現有關節的形狀。

Before you put the fingers in the positions you want, crease them all at a 90 degree angle. Even though real fingers bend in two places, I simplify it to one in the middle of the finger. Even if your fingers are going to be posed straight, creasing them first and pulling it straight again will give the illusion of a knuckle.

**11**

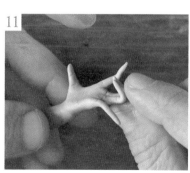

如照片所示，所有手指先彎折出折痕後，就可以擺出逼真的姿勢。

Once you've creased them all, you can perfect the pose.

用美工刀切除多餘的末端。

Cut off any excess at the end using your utility knife.

準備手臂用的銅線。將想要當成手肘的部分，彎折成圓弧的90度角後，再慢慢插入黏土手臂中。請注意，不要將銅線穿到手掌，只要穿至手腕和手掌的連接部分。穿入銅線時，即便手臂變形也沒關係。

Take your arm wire and bend it at a soft 90 degree angle where you want the elbow. Now very gently, insert the wire into the clay arm. The wire should not poke into the hand. Try to end it right where the wrist meets the palm and keep the hand untouched. It's okay if the arm gets misshapen as you work in the wire.

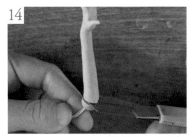

最好可以看到金屬線彎曲的部分（手肘部分），所以切除多餘的黏土。

You want the full bend of the wire to be visible so cut off any excess clay that covers it.

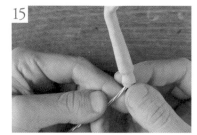

用自己的手指將黏土手臂根部該段，捏得比周圍細一些。

Press an indent at the base of the arm with your fingers.

最後在手臂表面塗上薄薄的消毒酒精使表面平滑。

And smooth the arm one last time with a thin layer of alcohol.

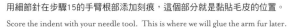

用細節針在步驟15的手臂根部添加刻痕，這個部分就是黏貼毛皮的位置。

Score the indent with your needle tool. This is where we will glue the arm fur later.

另一隻手臂也用相同方法製作。兩隻手都完成後放在耐熱容器或烤盤上燒製。燒製時間和溫度視黏土狀況而有不同，所以請確認外包裝或說明書。

Then repeat the process for the next hand. When your hands are done, place them on an oven safe plate or baking pan and bake them. Look at the instructions on the box of clay you are using and determine from that.

## 腳部塑造

將軟陶黏土（商品名稱：Super Sculpey）和彈性軟陶黏土（商品名稱：Cosclay）以1：1的比例混合做成腳。也可以用硬質的軟陶黏土代替一般的軟陶黏土。軟陶黏土具有足夠支撐身體的硬度，而彈性軟陶黏土具有柔軟度，所以不容易損壞。

I use a 1 to 1 mixture of Super Sculpey and Cosclay for the paws. You can substitute the beige Sculpey for Super Sculpey Firm if you prefer. The Super Sculpey will keep the legs rigid enough to support the body, and the Cosclay will keep the legs more flexible and resistant to breaking.

將兩種黏土完全混合揉捏成細長條狀，宛如高爾夫球桿的形狀。建議先畫出腳的完成圖，再比照捏塑。玩偶的腳是由這4條如高爾夫球桿的黏土組合而成。將2條形狀相同的黏土壓捏成一條。

Once your clays are thoroughly mixed, roll out a thin coil. Then shape it into a golf club shape. It might help you visualize having a paw diagram to compare your sculpt to. You will be sculpting four of these golf clubs for each paw making up the four toes. Sculpt an identical golf club. And press them together.

再做2條尺寸稍微小一點的高爾夫球桿狀的黏土。做出較小的條狀黏土後，在兩側稍微後面的位置夾住步驟2的黏土。腳的後側會形成空洞。

Sculpt the third and fourth toe slightly smaller, and position them a little further back on either side of the middle toes. You should have a hollow space behind your paw.

作出第5條細長狀黏土，將肉球部分捏成小小的三角形後，將黏土塞入步驟3的空洞，成了正中央的肉球。

Roll out a fifth coil but instead of a toe on the end, sculpt a little triangle at the base. Fit this coil into the hollow space to form the middle pad of the paw.

將接縫撫平接合。

Then smooth the leg seams out

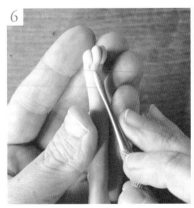 

添加腳趾輪廓。建議用矽膠型黏土刮刀劃出腳趾輪廓。

Add definition to the toes. The silicone wedge tool works best for separating the toes.

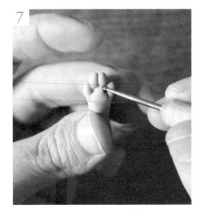

使用球狀黏土刮刀，加深肉球輪廓。

Use a small ball stylus to define the paw pad.

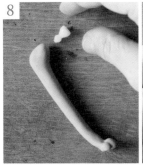

將直徑約1.5mm的 1 個娃娃骨架連接部件，插入腳的頂端。

Take one joint of your 3/32 armature. And press it into the top of the leg.

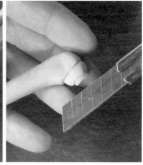

用美工刀刀刃的鈍角，在連接處刮去一圈黏土。

Use the blunt side of the utility knife to carve an indent around the ball joint.

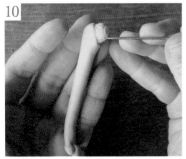

用細節針添加刻痕。

Score it with your needle tool.

另一隻腳也用相同方法製作。將兩隻腳左右對稱地放置在耐熱盤等器皿，和手臂一樣用烤箱燒製。

Repeat the process to make an identical leg. Place them on a plate symmetrical and opposite each other. Then bake them.

## 骨架和手腳結合

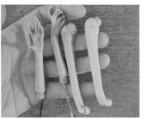

黏土冷卻後，用砂紙打磨至平滑。用浸泡過水的畫筆去除打磨產生的碎屑。為了避免塗裝剝落，不要忘記幫腳底打磨。

Once your hands and paws are cool, use your sandpaper to smooth out the imperfections. Remove all the sanding dust with a paintbrush dipped in water. Make sure to sand the bottom of the paw. This will keep the paint from scraping off.

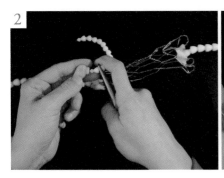

使用老虎鉗，將腳安裝在骨架上。用老虎鉗夾住骨架末端旁的連接部分（收窄的部分），並且用手指支撐，以免老虎鉗往後偏移。若有珠寶用的小型老虎鉗將方便作業。

Using your pliers, pop the leg onto your armature. Make sure your pliers grab the thin part of the joint behind the last one, and keep them from sliding back with your finger. I find thin jewelry pliers can be helpful at this stage if you have them available.

3

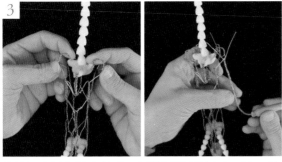

## Memo

彎折手肘時請彎折成圓弧的U字形。若彎折成銳角，會削弱銅線的強度。

When you bend the elbow, make sure you bend it in a soft U shape. Bending at a sharp angle will weaken the wires.

找出等同肩膀位置的紅色金屬線，讓手臂的銅線安裝在紅色、黃色、藍色金屬線交錯處。

Find the red wires that act as the shoulders. Place the arm wire at the point where the red, yellow, and blue wires meet.

4

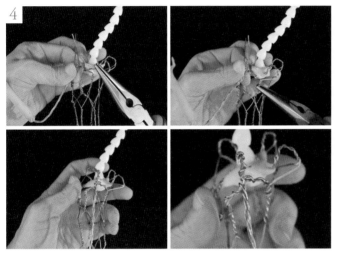

將手臂的3條銅線末端，從不同方向纏繞在細金屬網的金屬線。手臂的銅線分別纏繞在紅色、黃色、藍色金屬線。

Wrap each of your copper wire ends around the chicken wire in a different direction. Each copper wire wraps around each color, red, yellow, blue

5

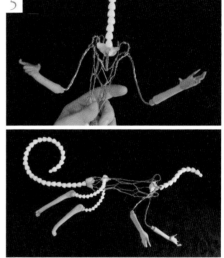

另一隻手臂也用相同方法安裝。

Then attach the other arm the same way on the other side.

6

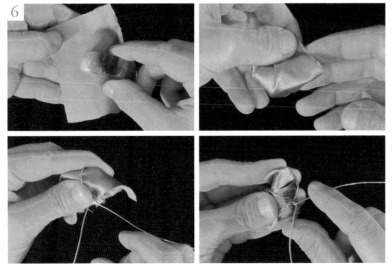

準備水晶心臟。創作者喜歡用碎毛皮布料或一般布料包裹水晶，但大家也可以直接安裝未經包裹的水晶。將銅線繞成一圈，從各個方向纏繞水晶。也可以使用花藝鐵絲。

Next we will add our crystal heart. I like to use a scrap of fur or fabric to wrap the crystal, but you can leave it bare. Loop some copper wire around the crystal and twist to keep it in place. Do this in many directions until the crystal stays secure. Floral wire works great for this too if you want to use that instead.

7

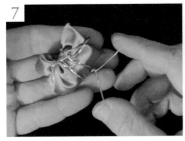

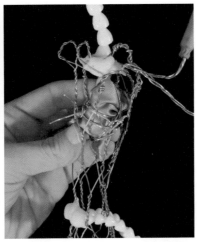

銅線的末端維持2條銅線分開、未固定的樣子。用這個部分捲繞在骨架上。拉開胸部周圍的金屬線，將水晶塞入細金屬網骨架的胸口。

Make sure to leave two prongs of wire free so you can secure it to the armature. Push the crystal into the chest of the chicken wire armature. You might have to stretch open the wires to fit it.

8

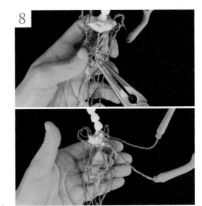

將銅線緊緊捲繞在骨架上，避免水晶鬆動。

Wrap the copper wire into the amature to secure the crystal in place.

## Memo

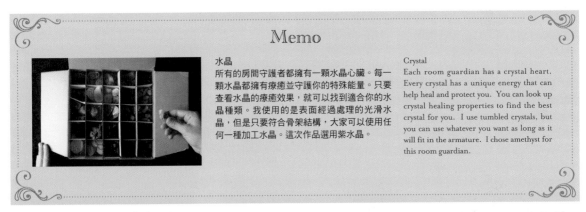

水晶

所有的房間守護者都擁有一顆水晶心臟。每一顆水晶都擁有療癒並守護你的特殊能量。只要查看水晶的療癒效果，就可以找到適合你的水晶種類。我使用的是表面經過處理的光滑水晶，但是只要符合骨架結構，大家可以使用任何一種加工水晶。這次作品選用紫水晶。

Crystal

Each room guardian has a crystal heart. Every crystal has a unique energy that can help heal and protect you. You can look up crystal healing properties to find the best crystal for you. I use tumbled crystals, but you can use whatever you want as long as it will fit in the armature. I chose amethyst for this room guardian.

9

在腳後側部分的細金屬網中放入小石頭。為了避免小石頭掉落，用鋁箔紙固定，這時稍微鬆一點沒關係。

Next, take your small stone and insert it into the chicken wire below the back legs. Make sure it does not stick out anywhere. It's okay if it is loose. We will keep it in place with aluminum foil.

10

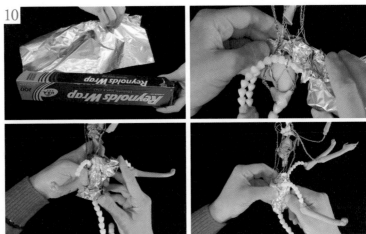

撕下長約12～13cm的鋁箔紙，揉皺後穿過細金屬網之間。往小石頭該側塞入，並且捲繞在金屬線周圍，以便固定小石頭。臀部周圍也捲繞上鋁箔紙，為了避免小石頭移動，塞滿間隙部分。

To start your foil, rip off a piece approximately 5 inches wide. Crinkle one end into a rough point and feed it through the chicken wire over the stone. Press it against the stone and wrap it around a wire to keep it in place. Then wrap the rest of it around the butt of the room guardian. Make sure to press it tightly into the holes to fill them and leave no space for the stone to wiggle.

11

重複步驟10，直到填滿臀部周圍間隙。只要鋁箔紙的密度夠大，用力緊握也不用擔心擠壞骨架。腳周圍也用鋁箔紙包裹後，用力推開擠壓鋁箔紙，以便妨礙腳的動作。

Repeat this step a few times to fill all the empty space in the butt. The goal is to get the foil dense enough to squeeze with all your might without collapsing. Make sure to rotate the legs around and push down any foil that obstructs their movement.

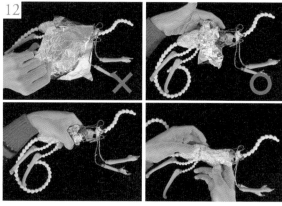

12

鋁箔紙不要像毯子一樣平面包裹住身體，而是以揉皺的狀態包裹捲繞。這樣才更能填補間隙。

Avoid wrapping the foil flatly around the body like a blanket. Instead, keep it rough and crinkly before pressing it down. That way the foil will fill in the holes more.

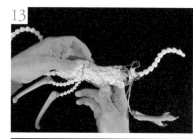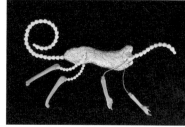

13

從上半身包裹至脖子的部分。請注意不要包裹到手臂。這個階段只要做出大概的形狀即可。身體可以用毛皮黏貼來添加分量，所以這個階段的作業要領在於維持細窄的身形。

Wrap all the way up to the neck but avoid wrapping the arms. The anatomy doesn't have to be perfect as long as you have the general shape of what the animal would look like without fur. It's best to keep this stage as thin as possible because the fur adds much more girth.

14

裁下長約10cm的鋁箔紙，揉皺後捲繞在手臂的銅線周圍。擠壓成人參的形狀。為了方便活動，手肘和肩膀周圍的銅線不要包裹鋁箔紙。

For the arms, rip off a 4 inch strip, and crinkle it around the copper wire of the arm. Squeeze it into a carrot shape. Keep the elbow and the shoulder wire free of foil for optimum movement.

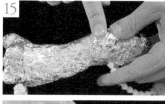

15

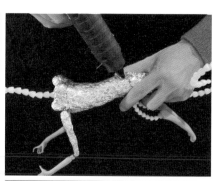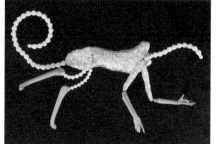

尋找剩下的間隙滴入熱熔膠後，塞入鋁箔紙碎屑填補間隙。

Find any residual holes in the foil. Add a drop of hot glue. And press a small piece of foil into the glue until the hole is filled.

16

手邊若沒有較薄的鋪棉就分成一半使用，並且塗上熱熔膠固定鋪棉的一邊，再橫向覆蓋在臀部。熱熔膠確定冷卻後才用手按壓固定，以免被熱熔膠燙傷。

Now take a strip of quilt batting. Peel it in half, unless you already have thin batting. Glue a line across the rump. Stick one end of your batting to the glue. Make sure to wait until the glue is cool enough not to burn your hand and press it onto the glue.

17

捲起約10cm的鋪棉黏在大腿後側。於表面裹上另一層鋪棉，並且用熱熔槍黏接在大腿內側。持續包覆整隻大腿。鋪棉和黏土交接的膝蓋部分，盡量將鋪棉收得細窄。剪去多餘的鋪棉，並且將邊緣黏貼固定。

Now take about 4 inches of excess batting and roll it up. Place this at the back of the thigh. Wrap the batting around it once and glue it along the inner thigh. Wrap the rest of the leg. Make sure to keep is as thin as possible at the knee where the batting meets the clay. Pull off any excess and glue the end down.

18

用織線綁縛大腿周圍。將織線捲繞在大腿周圍，纏繞出形狀。

Tie your yarn around the thigh. Wrap it around the leg to tighten the form.

19

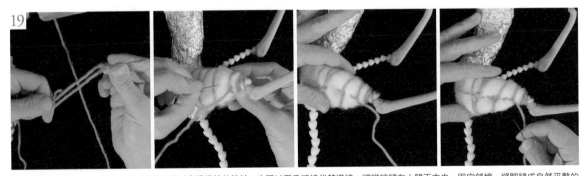

將織線纏繞至膝蓋後，穿過縫針。若沒有可以穿過織線的縫針，也可以用手縫線代替織線。將織線縫在大腿正中央。固定鋪棉，將腳縫成自然平整的形狀。

When you have thoroughly wrapped up the knee, thread the yarn through a needle. Optionally you can just use a needle and thread if you cannot fit the yarn through any of your needles. Then sew a line up and back down the center of your thigh. This will keep the batting in place and flatten the leg into a more natural animal shape.

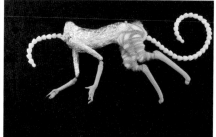

將多餘的織線纏繞整隻腳後，末端用熱溶膠槍塗膠固定。另一隻腳也用相同方法處理。

Wrap the leg all the way up with the excess and glue off the end. Repeat the process on the other side.

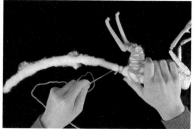

用熱溶膠槍將鋪棉黏貼在尾巴根部。用鋪棉包覆整條尾巴的娃娃骨架，從根部往末端，再從末端捲繞至根部。將根部包裹得比末端細。將織線纏繞住整條尾巴，並且固定在尾巴末端後，骨架即完成。

Get another strip of batting and glue a spot at the base of the tail. Wrap the batting around the plastic armature to the tip and back down. You want the base to be thinner than the end. Wrap the tail with yarn, and tie off the end. The armature is finished.

# 頭部塑造

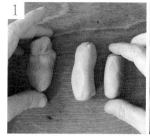

準備的 3 種黏土都用來製作頭部。建議軟陶土（商品名稱：Super Sculpey）使用量要多於彈性軟陶黏土（商品名稱：Cosclay），才比較容易雕刻，但是也可依照個人喜好調配。要注意的是，如果只使用軟陶土會比較容易破裂。黏土要充分混合，要混合出比預期需求量多的黏土。先做出狐狸口鼻部的大概形狀。捏出 2 個三角形當成耳朵。盡量左右對稱。

I mix all three clays for my head sculpt, but I use more Super Sculpey than Cosclay. I recommend more Sculpey than Cosclay for the best sculpting results, but feel free to mix your preferred consistency. Keep in mind that pure Super Sculpey will mean it will be more fragile. Knead your clays thoroughly together. Always mix a little more than you think you need. Then begin sculpting a rough muzzle shape for your fox. Pull out two triangles for the ears, and keep your sculpt as symmetrical as possible.

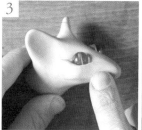

將 2 顆玻璃珠（直徑 8 mm）壓進黏土中。在眼睛的下面添加新月形黏土做出臉頰。

Press your glass beads into the sculpt. These are 8mm beads. Sculpt a crescent shape. Place it on the lower part of the eye to form the cheek.

將邊緣撫平。另一側臉頰也用相同方法製作。

Smooth the edges. Repeat on the other side.

**4**

眼睛上方也添加新月形黏土，做成眉毛。調整至表面平滑。

Add another crescent to the top of the eye to create the brow. Smooth it out.

**5**

將一片黏土添加在鼻子上方。

Add a strip of clay over the nose.

**6**

用手指將耳朵上部邊緣朝下轉，將耳朵下部邊緣往上挺，使整隻耳朵朝前。

Curl the top edge of the ears downwards with your finger. Then pull the bottom edge of the ear forward so the ears are facing the front

**7**

在耳朵背後添加圓形黏土，增加分量。

Add a circle of clay to the back of the ears to add some dimension to them.

**8**

添加新月形黏土，從兩側的側臉黏土至耳朵前面。利用工具使添加的黏土平滑接合。

Add a crescent of clay to each side of the face in front of the ears. Use a tool to smooth it.

**9**

在眉毛添加圓形黏土。利用工具使黏土平滑接合。

Add a circle of clay over each eyebrow. Smooth the edges with a tool.

**10**

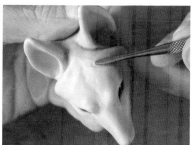

在額頭添加 2 片黏土，並且使表面平滑接合。

Add two pieces of clay to the forehead and smooth those out.

11

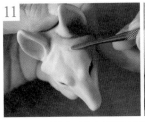 

使用塑形刀劃出顱骨的分割部分和眼睛後側的線條。

Use your wax carver to accentuate the split of the skull and the lines behind the eyes.

12

在口鼻部的兩側添加薄薄的橢圓形黏土。

Add a very thin oval to the sides of the muzzle.

13

添加小球狀黏土做成淚溝，用細節針按壓出平滑凹槽。用球狀黏土刮刀（小），加深淚溝輪廓。

Add a tiny ball of clay to the tear duct. Smooth it over the gap of the clay and the eye with your needle tool. Then use a small ball stylus to define the tear duct.

14

在耳朵內添加彎曲的條狀黏土，用球狀黏土刮刀（大）使黏土平滑接合。

Next add a curved coil into the ear, and smooth it with a large ball stylus.

15

 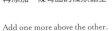

再添加一條彎曲的條狀黏土。

Add one more above the other.

16

  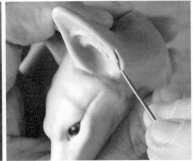

捏開耳朵下部的外廓，用細節針在外廓劃出分割線。

Pinch the bottom outer edge of the ear, and split the pinched edge with your needle tool.

**17**

用塑形刀使邊緣平滑後，塗上消毒酒精使耳朵平滑。另一隻耳朵也用相同方法製作。

Use your wax carver to smooth out two flaps. Then smooth the ear with some alcohol. Repeat the process on the other ear.

**18**

用美工刀切除脖子多餘部分。為了添加毛皮，一定要在耳朵後面保留約1.2～1.3cm。將1個直徑約3mm的娃娃骨架連接部件插入黏土。請慢慢小心押入，不要使頭部變形。

Use your utility knife to cut off excess on the neck. You should leave about ½ inch behind the ears to attach the fur later. Take one joint of your ⅛ armature and insert it into the clay. Push slowly and carefully to avoid messing up the sculpt.

**19**

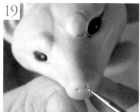

用球狀黏土刮刀（小）挖出鼻孔，用細節針在鼻孔添加橫向刻痕。

Poke two nostrils with your small ball stylus. Create the sides of the nostrils with your needle tool.

**20**

塑形刀押出鼻子下部，用細節針在鼻子中央添加刻痕。

Press in the bottom with your wax carver, and add the center line with the needle tool.

**21**

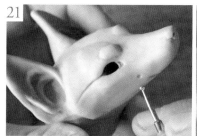

嘴巴末端用球狀黏土刮刀標記印記。用細節針劃出和印記相連的嘴巴輪廓線。

Mark the points where you want the corners of the mouth with your ball stylus. Then connect the dots with your needle tool to make the mouth.

**22**

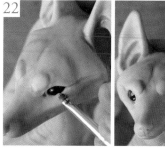

用少量消毒酒精將眼睛擦乾淨。在做出毛流質感前，請確認整個頭部的所有部位是否左右對稱，並且呈平滑表面。

Clean up the eyes with a little alcohol. Make sure everything looks symmetrical and smooth before you begin the fur texture.

## Memo

在營造毛流質感的作業中會使用到細節針。請盡量在靠近黏土的角度用針。照片中的上方是從低角度用針，下方是從高角度用針。若從高角度用針，可看出刮痕較深、也較紊亂。

Use your needle tool for your fur texture. Try to angle your tool as close to the sculpt as possible. The top is an example of fur texture with a low angle, and the bottom is an example of fur texture at a high angle. You can see it looks more scratched and messy.

從眼睛周圍開始營造毛流質感的作業,再往下和往後描繪。利用毛流質感可以加深眼睛周圍的輪廓。眼睛和鼻子周圍的毛流運筆較短,越靠近脖子拉得越長。

Start your texture at the eyes and work your way down and back. You can use the fur texture to define the line around the eye. Keep your strokes small near the eyes and nose and gradually get longer as you near the neck.

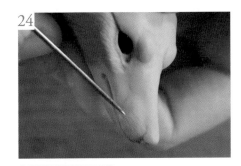

鼻子周圍的運筆又短又淺。

Keep the strokes especially small and shallow at the nose.

添加幾列較深的點表現出鬍鬚的毛囊。側面完成後從鼻子往中央添加毛流。

Then add a few rows of deeper dots to mimic whisker follicles. Once you've finished the sides, work your way up the center from the nose.

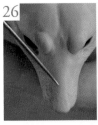

將鼻子上方的毛流描繪成扇狀展開的樣子,就可以和側面的毛流自然交會。劃至額頭後,從中心線往外側添加毛流。

Fan your strokes so they blend smoothly into the fur on the sides. When you get to the forehead, you can work your way out from the center line.

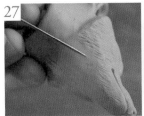

從嘴巴往脖子根部劃出毛流。

Work your way from the mouth to the neck from the bottom.

毛流質感描繪完成後,一邊塗上消毒酒精,一邊柔化毛流線條使邊緣平滑。

When the texture is finished, give your fox a light wash of alcohol to soften the lines and smooth any rough edges.

最後用細節針從脖子往兩耳劃出刻痕。

Lastly, score the neck and between the ears with your needle tool.

將頭部放入烤箱燒製,塗裝的事前準備即完成。

Bake your sculpt in the oven, and then it's ready for paint.

# 頭部塗裝

## 工具和材料

①壓克力顏料（高級）
②壓克力顏料（一般）
③極細畫筆
④各種粗細的畫筆
（舊的）
⑤消光透明保護漆

⑥防毒面罩
⑦光澤透明保護漆
⑧塑膠杯
⑨調色盤
⑩配合顏料色調的毛皮
碎布

①Golden Brand Paints
②Cheap Craft Store Acrylic
Paint
③Detail Brush
④Old Brushes in Varying
Sizes
⑤Mr. Super Clear UV Cut
Flat Spray

⑥Respirator
⑦Clear Gloss Varnish
⑧Cup
⑨Pallet
⑩Fur scraps to match paint
color

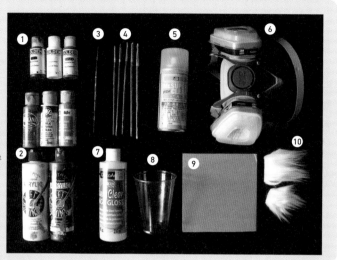

## 製作步驟

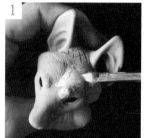

在頭部平均塗上一層薄薄的壓克力顏料（高級）Titanium white。不要塗到眼睛和耳朵內側。這個階段大致塗色即可。

Start with your golden brand Titanium white. Cover the head in thin even coat. Try not to paint the eyes and inner ears, but it's okay if you get a little messy

塗料乾了之後，在頭部下面塗上帶灰色調的白色。塗裝的色調選擇要配合該部位毛皮的色調。乾了之後，再塗上一層壓克力顏料（高級）Titanium white。

Once that coat is dry, I paint the underside of the head with some off white. This matches the tone of the fur I will be using for the front of the neck more, but you can judge the best paint to use based on your own fur choices. Then finish the rest of the head with another coat of Golden Brand Titanium White.

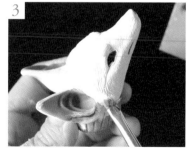

用壓克力顏料（高級）塗滿底色之後，薄薄塗上白色壓克力顏料（一般）。

Once you have a solid coat of your base color of Golden Brand paint, add a thin coat of the cheap white acrylic .

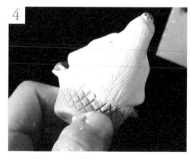

在頭部的下面塗上帶灰色調的白色。不斷重疊上色直到底色完成。

As well as more off white on the underside. Add as many of these coats as you need to get a solid color base.

### Memo

若眼睛沾到顏料，用沾濕的乾淨畫筆擦除。

Remove any paint on the eye with a clean wet brush.

混合黑色和白色的壓克力顏料調成淡淡的灰色，塗在眼睛周圍，並且描繪出頭和眉毛的輪廓。

Mix some black and white acrylic into light gray and paint a thin layer around the eyes. Use it to contour the head and brows.

5

6

同樣也畫出鼻子和嘴巴周圍的輪廓。

And do the same around the nose and mouth.

7

使用極細的畫筆，用黑色顏料畫出眼線。

Use your detail brush to line the eye in black.

8

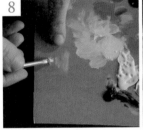

用乾的畫筆沾取白色顏料，淡化灰色部分的色調。用乾的畫筆先在調色盤上試塗，調至如粉筆般的柔和筆觸後，再塗刷上色。這個技法稱為乾刷。

Dry brush with white paint to soften the gray parts. The brush should be leaving soft chalk-like strokes on the pallet before you paint onto the face.

9

一邊對照左右兩邊一邊塗色。重複乾刷上色直到塗至理想的色調。

See the difference between the sides. Continue dry brushing until you achieve your desired softness.

10

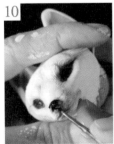

使用極細畫筆，用黑色顏料為鼻子和嘴巴塗色。

Paint the nose and mouth black with your detail brush.

11

在嘴巴的黑色顏料完全乾燥之前，用乾刷在嘴巴下部邊緣輕輕塗上白色，使白色和黑色部分混合，呈現出柔和的灰色漸層。

Before the black in the mouth is fully dry, dry brush some white along the bottom edge. It should blend with the black to add a soft gray gradient from the black to the white.

12

在鼻子周圍乾刷一層薄薄的白色，柔化邊緣。

Dry brush white around the nose to soften the edges.

用較小的畫筆在眉毛塗上黑點後，用乾刷塗上白色柔化色調。

With a small brush add black dots on the eyebrows. Soften them with a white dry brush.

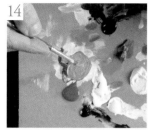

為了讓耳朵內側呈現自然的粉紅色，混合焦橙色、黑色、白色顏料。調成自然的粉紅色後，厚塗在耳朵內側。

For the inner ears, mix burnt orange, black, and white until you have a natural pink color. Paint the inside of the ears with a thick coat of this pink.

趁顏料未乾之際，在最深處添加少量黑色。黑色顏料如漩渦般往外延展抹開，混合成漸層色調。

While the paint is still wet, add a small amount of black to the innermost part of the ear and swirl it outwards so it mixes into the pink to form a gradient.

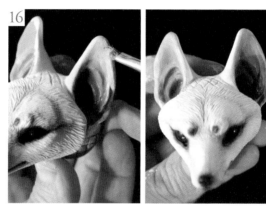

沿著耳朵邊緣塗上白色顏料，形成淡淡的粉紅色。

Then add some white along the edges until it forms a light pink blend.

顏料乾燥後，用乾刷塗上白色突顯邊緣。

When the paint is dry, refine the edges with some white dry brushing.

用乾刷在耳尖塗上黑色。

Add black tips on the ears with black dry brush.

為眼睛上色。將群青藍和白色混合成中藍色。顏料太濃時，添加少量的水稀釋。用極細畫筆塗出虹膜的顏色，並且保留瞳孔的部分。顏料過多時，塗裝表面會凹凸不平，還請留意。

To begin the eyes, mix some ultramarine blue with white until you have a medium blue. You can thin your paint with a little water if it is too thick. With your detail brush, paint the iris and leave out the pupil. Try not to overwork the colors or the paint will get lumpy and uneven.

**20**

虹膜的顏料乾了之後，在瞳孔周圍的輪廓畫上極淡的藍色原色線條。

When the iris is dry, paint a very thin line of pure blue around the edge of the pupil.

**21**

在中藍色中混入少量白色後塗在虹膜外側。

Mix a little more white into your medium blue and use this to paint the outer part of the iris.

**22**

沿著虹膜外側邊緣，畫上白色的細線條。

Then add a thin line of pure white along the outermost edge of the iris.

**23**

最後修飾時用耳朵塗色使用的粉紅色，塗在淚溝內側。頭部塗裝完成，即可準備塗上保護漆。

As a final touch, I like to add a little of the pink from the ears onto the inner part of the tear duct. Then the head is ready to be sealed.

## Memo

關於虹膜的基本塗色方法

靠近瞳孔的部分為暗色調，外側輪廓塗淡淡的顏色。我上色的方法大多是從中間色調開始上色，再慢慢塗上調淡的顏色和加深的顏色。顏料的透明度較高，若塗在暗色部分無法顯色時，請先將虹膜整體塗上白色，再於其上塗色，就會呈現鮮明的色調。

A general rule when painting irises is to paint the darkest color closest to the pupil and the lightest color on the outer edges. I like to start with my medium color and add the light and dark to it. If your paint is transparent and does not show up very well on the dark background, paint the whole iris white first and add your color on top of that. That will make the color more vivid.

## 手腳塗裝

**1**

從手臂到手腕塗上白色壓克力顏料（高級）。

Paint the arms up to the wrist with a coat of Golden Brand White.

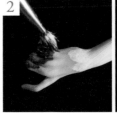

**2**

不要清除白色顏料，而是用畫筆直接沾取黑色壓克力顏料（高級），從指尖往上部塗色。因為和白色顏料混合，所以顏色越往上部抹開，黑色會越淡。一直塗到手臂正中央後，先將畫筆洗淨，去除黑色顏料。

Without washing out the white, dip your brush in the Golden Brand black paint. Start from the fingertips and work your way down. The black should get lighter and lighter as it mixes with the white. Once you get to the middle of the arm, wash the black out of your brush.

3

用畫筆沾取少量的白色顏料上色，和灰色部分混合形成漸層色調。這是打底階段，所以不需要描繪出完美的漸層效果。待第 1 層塗裝乾了之後，用乾刷塗上白色，讓色斑變得平滑。另一隻手臂也用相同方法塗色。

Add a little bit of white to your brush and blend the gray line into a smooth gradient. This is the undercoat so it doesn't have to be perfectly smooth. Once the first coat dries, you can smooth blemishes with a white dry brush. Repeat the process on the other arm.

4

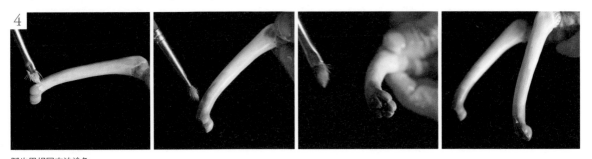

腳也用相同方法塗色。

Use this same method to paint the legs.

5

第 1 層乾了之後完全依照第 1 層的方法重疊塗上第 2 層顏色。第 2 層不使用高級壓克力顏料，使用一般壓克力顏料即可。

When the first coat has dried, paint a second coat exactly like the first only this time use the cheap acrylic paints instead of Golden Brand.

6

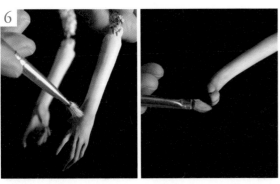

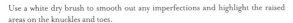

用乾刷塗上白色讓色斑變得平滑，在突起部分、關節和趾尖等處添加打亮。

Use a white dry brush to smooth out any imperfections and highlight the raised areas on the knuckles and toes.

7

用乾刷在指尖和趾尖塗上黑色變暗。

Use a black dry brush to darken the tips of the fingers and toes.

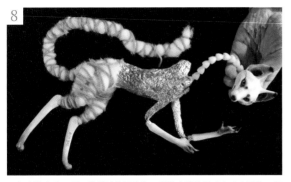

8

9

全部塗裝完成，即可準備塗上保護漆。雖然之後黏貼毛皮時會拆除，但事先將頭部安裝在脖子的娃娃骨架。

Now the paint is ready to be sealed. You can attach the head to the neck armature at this point, but you will have to remove it again for the fur.

將玩偶移到戶外或通風良好的空間，穿戴防毒面罩和防護手套。在塗裝表面平均噴上一層薄薄的保護漆（商品名稱：Mr. Super Clear）。將玩偶架好後，在距離約15～30cm處噴塗。也可以將玩偶放在塑膠板或厚紙板，在地面噴塗。要噴塗整個塗裝表面。

Move outside or to a well ventilated area, put your respirator on and wear protective gloves. Spray a thin even coat of Mr. Super Clear over all the painted parts of your room guardian. Keep the spray 6 to 12 inches away from the room guardian and move in a steady back and forth motion for an even coat. You can place the room guardian parts on a piece of plastic or cardboard on the ground. Just make sure you spray it from all angles.

10

保護漆乾了之後，用極細畫筆在眼睛、鼻子和嘴巴塗上光澤透明保護漆。使用光澤透明保護漆，會出現濕潤質感，讓玩偶看起來更生動。保護漆乾了之後，黏貼毛皮布料的事前準備即完成。

Once the Mr. Super Clear is dry, get your detail brush and clear gloss varnish. Coat the eyes, nose, and mouth with the gloss so they look wet and lifelike. Once the varnish is dry, the fox is ready for fur application.

## 毛皮布料的貼合

### 工具和材料

①縫針
②圓弧針
③珠針
④手縫線
⑤粉土筆
⑥刺繡彎剪刀
⑦裁縫剪刀
⑧打薄剪刀
⑨手工藝棉花
⑩針梳
⑪布用接著劑
⑫熱熔膠槍
⑬毛皮布料
（白色中長毛、用於脖子和腹部）
⑭毛皮布料
（白色長毛、用於脖子）
⑮毛皮布料
（白色長毛、用於尾巴）
⑯毛皮布料
（白色短毛、用於身體）
⑰毛皮布料
（灰色長毛、用於尾端）
⑱防塵面罩

① Straight Sewing Needles
② Curved Mattress Needle
③ Quilting Pins
④ Thread
⑤ Colored Pencil
⑥ Curved Embroidery Scissors
⑦ Sewing Shears
⑧ Pet Grooming Thinning Shears
⑨ Stuffing
⑩ Slicker Brush
⑪ Fabric Glue
⑫ Hot Glue Gun
⑬ White Medium Pile Fur for neck and belly
⑭ White Long Pile Fur for neck
⑮ White Long Pile Fur for tail
⑯ White Short Pile fur for body
⑰ Gray Long Pile fur for tail tip
⑱ Dust Mask

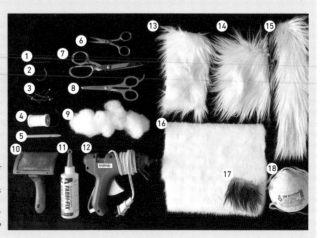

# 縫製方法
## A Note on Sewing

縫製玩偶時會使用「毛邊縫法」和「棒球縫法」兩種縫法。

I use two types of stitches when I am sewing a room guardian: The Blanket Stitch and the Baseball Stitch.

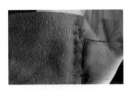

**毛邊縫法**
毛邊縫法是一般手工藝中相當廣泛使用的縫法，適合用於隱藏毛皮布料的接縫處，還可防止裁切邊緣綻開。這種縫法適用於將布料的毛皮面相對從反面縫合。

The Blanket Stitch is a very common stitch to use in crafting and the best way to hide fur seams and keep the edges from fraying. It can only be used when you are able to place the furry sides of your two fabrics together and sew from the back side.

**棒球縫法**
棒球縫法適合用於毛皮布料無法翻至反面，不能從反面縫合的時候。縫針往上又往下刺穿縫合，宛如棒球的縫線般而得此名。

The Baseball Stitch is used when you have to sew the fur from the front furry side and can't be flipped to the back. It is called the Baseball Stitch because you pierce the needle under and over like the stitches on a baseball.

## 毛邊縫法

**1**
將 2 片布料的毛皮面相對。

Place the two furry sides of your fabric together.

**2**
用縫針刺穿 2 片布料，收緊縫線至縫線打結處固定。

Pierce the needle through both sides. Pull it through until the knot catches.

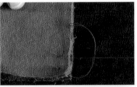

**3**
將縫針再次從相同方向刺穿，並且穿過縫線成圈的部分，接著重複這個步驟。

Pierce the needle through the fabric the same direction as before, but feed the needle through the loop. Repeat this stitch.

**4**
收緊縫線，讓接縫處緊密。

And pull it tight to seal the seam.

**5**
重複這個步驟直到接縫處完全閉合。

Repeat this until the whole seam is sewn.

**6**
用迴針縫收針後剪去多餘的縫線，所以使用毛邊縫法時會有 2 ～ 3 cm 往回縫的針腳。

To lock your stitch before you cut the thread, sew a few centimeters back up the seam with the blanket stitch

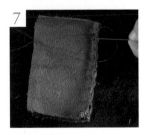

**7**
迴針縫的針距又小又密。

Keep the stitches small and close together

縫2～3針後，剪去縫線。在這次的範例中，為了方便展示，刻意縫成較寬的針距。實際製作玩偶時，會縫成較小又較密的針距。

Do that a few times and then cut the thread. These stitches are large for better visibility, but I use much smaller stitches closer to the seam on the actual room guardian

## 人造皮草的毛邊縫法

請注意不要將毛皮縫進接縫處，將毛皮面相對重疊。

When joining the furry sides together for the blanket stitch on faux fur you want to make sure to keep the fur out of your seam.

用縫針將所有的毛往內塞避免將毛縫進接縫處。

To do this, take a needle and brush it down the seam of your fur until all the loose hairs are tucked.

捏緊毛皮並且用珠針固定，避免位移。

Pinch it to keep the hairs tight and pin it to keep it in position.

## 棒球縫法

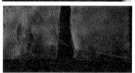

將縫針從布料反面往上穿出，將縫線收緊至縫線打結處。

Pierce one side from underneath and pull it out the top of the fabric. Pull it until the knot catches.

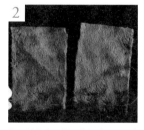

另一片布也一樣，將縫針從反面穿出後收緊縫線。

Pierce the needle through the opposite side the same as the first. Pull it tight.

就像穿鞋帶一樣，將針從反面往正面穿出，形成接縫處。

Repeat this back and forth down the seam like you are lacing a shoe.

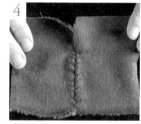

毛皮可完美遮蓋住接縫處。翻到反面就可以清楚看到接縫處。

The stitches are well hidden by the fur, so this shows you the stitches more clearly from behind.

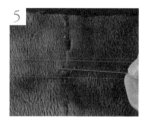

為了收針，往回縫幾公分，如同毛邊縫法一樣，將縫針穿過圈中。將縫針從反面往正面穿出，縫2～3次後剪去縫線。

To lock your baseball stitching, stitch a few centimeters back up the seam. Loop the needle through itself like the blanket stitch. Go back and forth a few times and then cut the thread.

## 人造皮草的棒球縫法

縫合時盡量將毛從接縫處挑出，避免將毛縫進接縫處，用手指按壓住毛，將縫針穿過毛皮布料之間。

To keep the fur out of your stitches while doing the baseball stitch, pull the fur away from the seam as often as possible to keep it clear. Press the fur down with your finger as you pierce your needle through the fur.

為了讓縫針穿過圈，先用手拉住，避免纏繞到毛皮。

Hold the loop as you feed the thread through the hole to keep it from getting tangled in the fur.

如上面的照片所示若有許多毛卡在接縫處,用縫針穿過接縫處,將纏繞成圈的毛挑出。

If you do get a lot of fur caught in your seam as shown above, poke your needle underneath the loop of tangled fur and pull it out of the stitches.

建議一邊縫一邊依照步驟 3 的方式處理。

It's best to do this as you go.

製作步驟 ……………………………………………………………………………………………………………

# 頭部貼上毛皮布料

長毛的毛皮布料放在脖子後面,用粉土筆在兩耳的正中央的位置,在布料標註記號。照片中脖子放在尺寸適中的毛皮布料上,但是也可以在尚未裁切、尺寸較大的毛皮布料上標註記號。

Take the long pile fur for the back of the neck and mark the midpoint between the ears with your pencil. You can do this directly from a large piece of your faux fur instead of having a small pre-cut piece.

在布料貼在兩頰的部分標註記號。

Mark the point on each side where the fabric reaches to the cheeks.

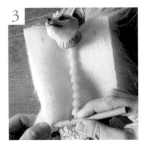

在布料貼在雙肩上部的部分標註記號。

Then mark where the fabric reaches the top of the shoulders.

在脖子正中央部分畫線,並且將脖子最上部等 5 個標記點之間畫線相連。雙肩之間的線條有點彎曲。

Draw a line down the center of the neck and then connect the five dots. Make sure to curve the line between the shoulders slightly.

用裁縫剪刀沿著輪廓線裁切。慢慢移動剪刀裁切,以免剪到另一面的毛。

Cut the shape out using your fabric shears. Make sure to use small snips close to the fabric backing to avoid cutting the fur off the backing.

將裁下的毛皮布料比對玩偶,確認是否符合大小身形。毛皮布料必須從頭部的切口覆蓋到未塗裝的部分。

Once your pattern is cut out, place it on the room guardian to make sure it fits. The fur should fit along the scored unpainted edges of the head.

測量毛皮布料前端覆蓋頭部的部分和胸部的寬度,並且決定脖子前側的毛皮布料形狀。再次確認測量的長度是否正確後,用裁縫剪刀裁切。

Measure the length between the edges of your fur on the head and chest. Use these measurements to make the pattern for the front of the neck on your medium pile fur. Double check your pattern with your cut fur to make sure the measurements are right. The pieces should match up nicely once they are cut out.

**8**

用珠針固定 2 片毛皮布料。珠針和接縫處呈垂直插入。毛若超出接縫處，就用珠針將毛往內塞回。用毛邊縫法從頭部開始縫合至下端後，再用迴針縫往回縫 2～3 cm。盡量縮小縫份。

Pin the two pieces together with your quilting pins perpendicular to the seam. Make sure no fur is poking out from the seam. You can use a quilting pin to brush in any stray fur. Beginning at the top where the head will be, use the blanket stitch to sew the seam together. When you reach the end, sew back up the seam a few centimeters to keep your stitches from coming out. The seam allowance should be very small.

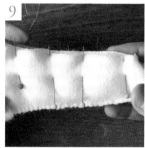

**9**

另一側也用珠針固定，整體如筒狀。另一側也同樣縫合。兩邊都縫合後翻回正面。

Pin the opposite side to form a tube, and use the same method to sew up this seam. When both sides are securely stitched, flip the tube inside out.

**10**

用老虎鉗拆除頭部。將步驟 9 的毛皮布料穿過脖子骨架後，重新安裝頭部。將毛皮布料拉至頭部有刻痕的部分。

Remove the head of the fox with your pliers. Place the fur tube over the neck armature, and reattach the head. Pull the fur over the scored part of the head.

**11**

用布用接著劑將毛皮布料黏貼在頭部。從兩耳的正中央黏貼。大概包覆住全部有刻痕的部分。

Starting in the midpoint between the ears, glue the fur to the head sculpt using your Fabri-Fix. This should cover most of the scoring on the head.

接著劑硬化後，準備長毛的毛皮碎布。使用刺繡彎刀剪下毛皮布料上的毛。確認剪下的毛流和脖子後面的毛流方向是否相同。捏起一撮毛後，從毛皮布料剪下，盡量讓切口平整，再用布用接著劑輕輕黏貼在耳後。用另一隻手指輕輕拍黏合。請注意，若太用力按壓於一點，手指會沾到接著劑而黏到毛，反而會拔下所有的毛。

Once the glue is set, take a scrap of your long pile fur, and cut the fur off the fabric backing with your embroidery scissors. Make sure the fur is pointing in the same direction as the fur on the back of the neck. Pinch the fur to pull it away from the backing, and try to keep the trimmed base as clean and smooth as possible. Apply Fabri-Fix glue along the seam behind the ear. Gently place your fur onto the glue, and tap alternating fingers repeatedly along the glue to press the fur down. Avoid pressing too hard in one place or the glue will stick to your finger and pull off the fur.

等待數分鐘接著劑硬化後，用針梳刷除多餘的毛。

When the glue sets in a few minutes, use your slicker brush to remove the excess fur.

若從毛皮表面仍可看出明顯的接縫就再補上毛。另一側耳後也用相同的方法黏上毛。

If the seam shows through your fur, just add a second layer exactly like the first. I used two layers on this fox. Repeat the process on the other side.

從毛皮布料剪下一塊毛，配合脖子毛流的方向，用布用接著劑黏貼在兩耳之間。

Cut a clump of fur off the backing. Apply glue to the space between the ears. Press the fur into the glue to match the fur direction of the neck.

從短毛布料剪下毛後，用布用接著劑黏貼在耳朵內側的上部。

For the inner ears, cut a thin line of your short white fur off the backing. Add a thin line of glue along the top inner ear. Gently press the fur into the glue.

接著劑乾了之後，將黏貼的毛用刺繡彎剪刀修整至耳朵正中央的長度。

Once the glue is set, use your embroidery scissors to trim the fur to the middle of the ear.

耳朵內側的下部也黏貼上毛。左右毛之間會有一點縫隙。之後會在這個部分塗色點綴。另一側耳朵也用相同方法處理。

Add glue to the bottom of the inner ear. There should be a visible gap between the two sides. We will accentuate it later with paint. Repeat on the other ear.

41

**19**

 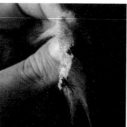  

準備灰色長毛的毛皮碎布。這次為了突顯臉頰和脖子的毛，使用暗色調。將剪成細長的新月形，用布用接著劑從耳朵上部黏貼至嘴巴下部。另一側也用相同的方法黏貼。

Take a scrap of your gray fur. I like to use dark fur for this step because it will create definition between the cheek and the neck fur. Cut out a thin crescent shape. Fit it onto the cheek. It should reach the top of the ear to just below the mouth. Cover the fabric backing with glue. Repeat on the other side.

**20**

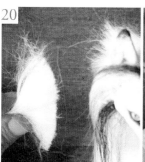   

**21**

從白色長毛布料剪下線狀的毛，再用布用接著劑黏貼在步驟19的毛皮接縫。接著劑乾了之後，用針梳刷除多餘的毛。另一側也用相同的方法處理。

Cut a line of white long pile fur off the backing. Glue along the seam of the gray fur. Press the fur into the glue. Brush out the excess once the glue sets. Repeat on the other side.

剪下的毛束沿著臉下方的形狀彎曲，沿著接縫用布用接著劑黏貼。接著劑乾了之後，用針梳梳理。

Curve your fur beforehand to fit the shape of the head. Glue along the bottom seam of the face. Press the fur into the glue and brush it out once it's set.

**22**

為了可看出下巴的鬍鬚，將從毛皮布料剪下的毛用接著劑黏貼在下巴。

Apply fur to the bottom of the face. This will make the chin slope naturally into the fur. It should look like a beard.

**23**

為了讓接縫處更平順，在臉的下面再黏一層毛，並且用刺繡彎剪刀修剪。剪刀要配合毛流修整，以免末梢出現參差不齊的樣子。

Add more layers of fur to the bottom of the face to smooth any uneven looking seams and use your embroidery scissors to trim the hair. Make sure you trim parallel to the fur direction so you don't get any choppy blunt edges.

## Shadowmask Daemon Vitale

「守護神Vitale」為創作者設計的一種原創幻獸，以樹脂
鑄模翻模製作。耳朵有安裝球狀關節，所以可以變換耳朵
的姿勢。

Daemon Vitale are one of my original species. This one is made with
cast resin that I molded from an original sculpt. Unlike the winter fox,
the ears are ball jointed and can be posed.

## The Voids

「Void」也是原創幻獸的一種，每隻都
有不同的身體造型和姿勢。他們沒有
臉，而是用天鵝絨覆蓋在臉部，所以看
起來像有個黑洞。造型亮點是利用層次
重疊的技法在脖子的毛皮添加羽毛。

Voids are also one of my original species. Each
one has a different body with differing posability.
The faces are hollow and lined with velvet so
they look like black holes. Up close you can see
I have added feathers into their neck fur using a
layering technique.

**24**

剪下一縷黑色和灰色的毛，用布用接著劑沿著眼線黏貼。為了讓毛流平順，捏去多餘的毛。另一側也用相同的方法處理。

Cut a small tuft of black and gray fur. Apply a small dot of glue along the dark eye line and press the gray tuft into the glue. Pinch out the excess until the line blends smoothly with the fur, and repeat on the other side.

**25**

在眼線上部的接縫處補黏上少許的白色長毛。最後在臉的上部中央間隙和其他所需之處也補黏一些毛。

Add more long white fur to the seam above the eye line. Then add the final fur to the middle gap and any other parts that need it.

**26**

用針梳刷蓬臉頰的毛。修整太長或遮蓋住耳朵的毛。兩耳之間的毛，順著臉部毛流修整。

Fluff the cheek fur with your brush and trim the areas that obscure the ears. Then trim the hair between the ears so it blends into the short fur detail on the face sculpt.

**27**

在脖子一點點塞入小片的手工藝棉花，直到包住整個娃娃骨架的周圍。若塞得太滿，會限制脖子的動作，所以少量即可。

Take your polyfill and pull it into small fluffy bits. Stuff the neck one bit at a time. Make sure you get all the way around the plastic armature. The neck should be only lightly filled. Avoid stuffing it too densely or you will restrict the neck movement.

## 身體貼上毛皮布料

**1**

將毛皮布料剪成大小適中的方形。將玩偶放在布料的一角，手臂和身體擺成45度角的位置。手肘稍微超出毛皮布料的一邊，布料的下端稍微超過膝蓋。在背後的位置標示中心線的記號，在雙肩的位置標示頂端線條的記號。

To cut out the right size square of fur, lay your fox in a corner and place the arm at a 45 degree angle from the body. The edge of the fur should extend slightly past the elbow. The bottom of the fur should extend slightly past the hock. Mark where the back was for the center line and where the shoulders were for the top line.

將玩偶轉向另一側，也畫出相同的線條。將畫出的線條連接，就會構成類似方形的形狀，再用裁縫剪刀沿著輪廓線剪裁。

Roll the fox over to mark the lines on the other side. You should have a rough square shape with a center line. Cut it out with your fabric shears.

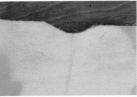

將中心線的上端和脖子後面的毛皮下端相連，可描繪出一個微彎的新月形，將銳角處剪成圓形裁去。

At the top of your center line, draw a shallow crescent shape roughly matching the curve of the bottom of fur on the back of the neck. Cut it out and round out the corners.

剪去新月形的部分放置在脖子根部，將手縫線穿過縫針。

Lay your body fur over the head of your room guardian with the crescent cut positioned at the base of the neck and thread your straight needle.

這個角度不容易用珠針固定毛皮布料，所以用毛邊縫法從正中央開始縫合至脖子前面的接縫處。另一側也用相同的方法縫合。

Since the angle of the fur is too awkward to pin, start your stitching in the center and work your way out. Use the blanket stitch and stop when you reach the front neck seam. Repeat on the other side.

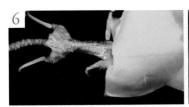

用熱熔膠槍在背後的中心線塗上黏膠。確認毛皮的中心線和塗上黏膠的線條是否一致後，按壓黏緊至黏膠冷卻。

Use your hot glue gun to glue a line down the center of the back. Make sure the center line of the fur matches up with the line of glue, and press it down firmly as it cools.

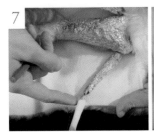

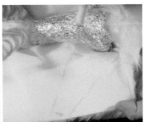

將手臂擺放成45度角，沿著上臂上側畫線，從脖子接縫處畫至手肘。將毛皮布料捲覆手臂，畫出下側線條，並且畫至和上臂上側的交會處後，裁切下來。

Lay the arm out at a 45 degree angle on the fur, and draw a line along the top of the arm from the neck seam to the elbow. Roughly wrap the fur around the arm and draw a line where the bottom meets the top line. Then cut it out.

毛皮邊緣塗上布用接著劑，中心線塗上熱熔膠。一邊注意不要產生皺褶，一邊將毛皮按壓在手臂。將塗上布用接著劑的邊緣部分，黏貼在手臂部件劃出刻痕的部分。

Glue the top edge of the arm fur with Fabri-Fix and glue down the middle of the fur with hot glue. Press the fur onto the arm, making sure there are no wrinkles. Then affix the Fabri-fix edge onto the scored part of the arm sculpt.

 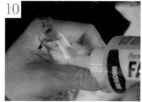

固定毛皮、彎曲手臂，即便接著劑乾了之後，也不會影響手臂動作。

Hold the fur firmly as it sets and bend the arm. This will ensure that the glue dries without restricting movement.

接著劑乾了之後在毛皮接縫處塗上布用接著劑，均勻緊密黏貼在手臂，以便遮蔽接縫處。請注意不要按壓得太用力。

Once the glue sets, glue a thin line of Fabri-fix along the fur seam. Lightly press the arm fur evenly into the glue to hide the seam. You only need a thin layer of fur to stick so don't press to aggressively.

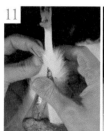  

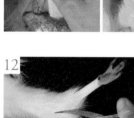

接著劑乾了之後，用棒球縫法從手肘縫合至腋下。縫合後用針梳梳理，使接縫處變得不明顯。

Once the glue is set, thread a straight needle and use the baseball stitch starting at the elbow and working down to the armpit. Brush it out with your slicker brush. The seam should be invisible.

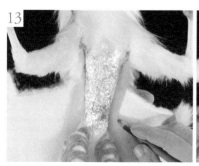 

用刺繡彎剪刀修整毛。彎曲手臂用剪刀稍微修整手肘彎曲的地方。手臂根部的毛較長，越往前面毛越短，就會使手臂線條平順纖細。修整毛流使身體和手臂部件的毛流自然相連，另一隻手臂也用相同的方法處理。

Use your embroidery scissors to trim the fur. Bend the arm to cut a sharp corner at the elbow. Keep the fur longer at the bottom and shorter at the top. You want the arm to look gently tapered, so when you're done the fur should blend smoothly into the arm sculpt. Repeat the same process on the opposite arm.

將毛皮黏貼在身體側面，在等同手臂～大腿上端的部分畫線後裁切。

Press the fur flush to the sides of the body. Draw a line from the arm to the top of the thigh on either side. Then cut the lines.

14

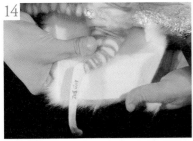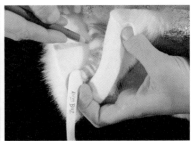

將大腿平放在毛皮布料上。將毛皮布料對摺包覆大腿上側，並且在毛皮布料重疊的邊界標註記號。

Flatten the fur against the thigh and hold it in place. Fold the fur over the top of the thigh, and mark the line where the fur meets.

15

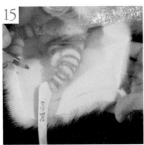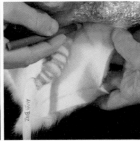

將包覆大腿上側所標註的邊界線相連至膝蓋處，並且畫出邊角，就形成包覆大腿的摺痕線條。依照線條剪裁確認布料是否可相對接合。

Connect those two lines at the hock. Then connect the top corner of the thigh with the corner of your line. Blue shows where this line will wrap onto the thigh. Cut your pattern out and double check that the seams meet up nicely.

16

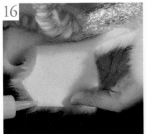

在大腿部分的毛皮布料下端塗上用布接著劑，在中心線塗上熱熔膠。將毛皮布料的兩邊相對包覆腳，但是不要產生皺褶。將塗上用布接著劑的部分，緊貼按壓在腳部件劃有刻痕的部分，直到接著劑乾燥。

Just like the arms, glue a line of Fabri-Fix at the base of your pattern, and a line of hot glue down the center. This time curving slightly with the leg. Then wrap the fur around the leg making sure the two sides meet up cleanly and there are no wrinkles. Hold the Fabri-Fix seam onto the scored part of the leg until the glue sets.

17

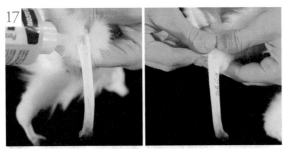

在毛皮布料的接縫處塗上用布接著劑，輕輕按壓毛皮布料。

Glue a thin line around the edge of the seam like you did the arms. Gently press the fur into the glue.

18

用棒球縫法從膝蓋縫合至臀部。將縫線成圈的部分往反方向拉，避免纏繞到毛。用迴針縫收針後，梳理接縫處。

Use the baseball stitch to sew from the hock to the butt. To keep your thread from getting tangled in fur, hold the loop as you feed it into the other side. Then to lock your stitch, sew back up the seam, cut the thread, and brush out the seam.

19

 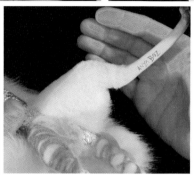

用刺繡彎剪刀修整出腳的形狀，將大腿內側的毛修剪成平整的短毛。用打薄剪刀修整，避免短毛和長毛出現明顯的層次差異。為了讓大腿形狀更明顯，在膝蓋保留一點毛。和手臂一樣，大腿邊緣的毛較短，根部的毛比較蓬鬆。另一隻腳也用相同的方法處理。

You can sculpt the shape you want for your leg with your trimming. Trim the inner thigh short to flatten the shape. Then use the thinning shears to blend the trimmed fur gradually into the longer fur. When you get to the knee, leave a little excess fur on the bend to give it a sharper shape. The fur on the upper thigh should be shorter and the lower thigh should be longer like the arm. Repeat on the opposite leg.

20

準備中長毛的布料，增加玩偶手腳的分量。將毛皮布料放在腹部鋁箔紙露出的部分，畫出輪廓線後，沿著線條剪裁。

Take your medium pile fur and spread the arms and legs of the fox. Place the fur over the foil on the belly, trace the edges, and cut it out.

21

  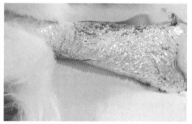

將毛皮重疊在左右的任一側大腿的上部。使用圓弧針以毛邊縫法從大腿邊緣縫合至肩膀。因為這個毛皮布料位於不容易固定的位置，為了不要位移而從此處開始縫合。

Choose a side and join the corresponding corner points right above the thigh. Since this fur can't easily be pinned, we will start sewing from here to keep the fur in the right position. Since this seam is at an awkward angle, the curved needle makes sewing much easier. Use the blanket stitch from the corner of the thigh up to the shoulder.

22

將毛皮摺起包覆腹部，為了避免位移在毛皮和骨架插入珠針固定。用圓弧針以棒球縫法從大腿邊緣縫合至肩膀。

Fold the fur over the belly and insert pins through the fur and into the armature to keep it in position. With your curved needle, use the baseball stitch from the corner of the thigh to the shoulder.

23

將手工藝棉花塞入脖子間隙等分量
不足的地方，並且用棒球縫法縫
合。因為這裡是施力擺出動作的地
方，所以要緊密縫合。

Check the remaining neck opening
and add extra stuffing if it feels hollow.
Stitch it closed with the baseball stitch.
Make sure you secure this seam very
well because posing the neck and arms
causes a lot of stress on these parts.

24

接著縫合髖部。用珠針固定毛皮布
料，以棒球縫法從大腿上方縫合至
尾巴根部。另一側也用相同的方法
處理。

Move on to the crotch next. Pin the fur
into position. Begin at the top of the
thigh and use the baseball stitch down to
the base of the tail. Repeat on the other
side.

25

用打薄剪刀修整毛流，接
近大腿部分的毛較短，越
接近胸部毛越長。

Use your thinning shears
to shorten the fur closest to
the thighs and gradually get
longer toward the chest.

26

將尾巴放在長毛布料上，裁
成可以包裹整條尾巴的大
小。將灰色毛皮布料放在尾
巴末端，決定長度後，配合
大小裁去白色毛皮布料多餘
的部分。將尾巴末端放在灰
色布料的部分。

Lay your tail out onto your
long pile fur. Cut out a strip
that will wrap around the width
of the tail. Place your gray fur
at the base of the strip and
determine how long you want
the tip to be. Mark the line
where you want the gray tip to
start, cut the excess white off,
and place the gray on the tip.

27

依照白毛布料的寬度，將灰色布料裁切成橢圓形。

Mark a U shape that fits to the width of the white and
cut it out. Make sure it fits perfectly together.

28

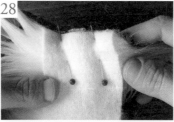

用珠針固定 2 片毛皮布料，再用毛邊縫法縫
合。

Pin them together. Use the blanket stitch with a
straight needle to sew up the seam.

大腿部分的毛皮布料超過尾巴時，將布料剪至尾巴根部。

If there is any excess fur on the tail from the thighs, just cut it off at the base of the tail.

將尾巴毛皮布料放在身體的尾巴根部，用圓弧針以毛邊縫法從正中央縫至尾巴側邊。另一側也用相同方法縫製。接縫處必須包覆住整個尾巴根部的周圍。

Place your tail fur at the base of the tail. Begin in the middle and use your curved needle to blanket stitch one side, then stitch the other side. This seam should stretch completely around the base of the tail.

將尾巴的毛皮布料正面相對縫合。為了不要讓毛捲入接縫處，可能會需要花一點工夫縫製，從尾巴末端到白色毛皮的接縫處，一邊將超出的毛往內壓，一邊用毛邊縫法縫合。

Now take your tail tip and fold it inside out. It will take some maneuvering to get the fur out of the seam but just start at the very tip and push the fur in as you work your way down. Begin sewing the seam with the blanket stitch. Stop when you reach the white fur seam.

先不要將縫線剪斷，若有需要利用工具將尾巴末端翻回正面。

Without cutting your thread, push the tip back right side out. You may need to use a tool.

將尾巴骨架塞入尾巴末端。在毛皮布料中央部分的各處塗上熱熔膠後，固定住尾巴骨架。

Push the tail armature into the tip of your fur. Glue several dots of glue down the center of the fur to keep it in place.

尚未縫合的部分以棒球縫法縫合。

Sew the rest of the tail using the baseball stitch.

**35**

梳理接縫處後即完成。

Brush out the seam. All the sewing is finished.

## 最後修飾

**1**

在最後修飾時塗上點綴的色彩。用水稀釋黑色壓克力顏料,塗在耳朵內側,也就是左右毛的間隙。

For the final touches, get some black acrylic and water to paint accents into the fur. Coat a small brush lightly with diluted black paint. Darken the gap between the fur on the inner ears.

**2**

同樣淡淡塗在肩膀之間後,用梳子刷開。若塗太厚會讓毛失去柔軟度,還請小心。若想讓顏色更暗,從表面薄塗重疊上色。

With a bigger brush, dab some diluted black across the shoulders. You don't want the paint to be too thick or it will affect the softness of the fur. Smooth it out with your brush. Add another layer if you want it darker.

**3**

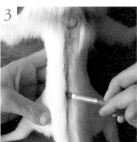 

在背後的線條也塗上顏色,並且用梳子將顏色刷開。

Add a line down the back. Brush it out.

**4**

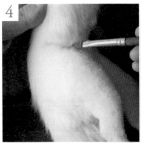 

將顏料稀釋至灰色後,塗在腳跟及腋下,再用梳子刷開。利用梳理完成沉穩的色調。至此作品即完成。

Dilute your black with water until it is a faint gray. Use it to accentuate the creases around the legs and brush it out. Do the same under the arms. It should be very subtle once it's brushed out. The Winter fox is finished.

# 鹿龍 Lilacoo

這位個人創作者運用羊毛氈和藝術玩偶的製作技法，使幻獸的外表兼具真實與虛幻的特色，逼真靈動，栩栩如生。
不論是腳的皺褶，還是毛流、爪子、牙齒的生長型態等，不放過任何部位細節的雕琢，對於生物樣貌的塑造傾注相當多的心力。
作品主要展示於Twitter、SNS等社群平台，未來也計畫舉辦個人的作品展。

[HP] lilacoo2.jimdofree.com
[Instagram] lilacoo_nf
[Twitter] @sweet_mokaketu

# 鹿龍的製作方法

工具和材料

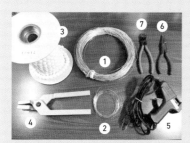

①金屬線
②細金屬線
③娃娃骨架
　（S尺寸）
④娃娃骨架專用鉗
⑤熱熔膠槍
⑥老虎鉗
⑦斜口鉗

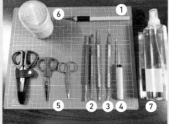
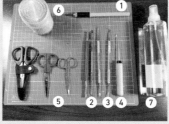

①工藝刀
②黏土刮刀
③塑形刀
④錐針
⑤剪刀
⑥牙籤
⑦噴霧瓶

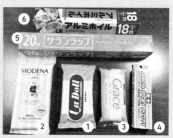

①石粉黏土
②樹脂黏土
③樹脂風黏土
④環氧樹脂
⑤保鮮膜
⑥鋁箔紙

①壓克力漆消光黑
②稀釋液
③壓克力顏料
④畫筆
⑤透明打底劑
⑥水性壓克力保護漆

①砂紙
②海綿砂紙

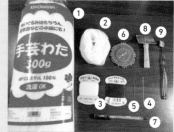

①手工藝棉花
②毛線
③手縫線
④縫針
⑤羊毛氈戳針
⑥珠針
⑦筆
⑧針梳
⑨牙刷

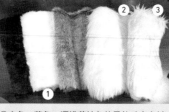

①UV燈
②樹脂液
③樹脂用表面修補液
④半圓球模具
⑤拋棄式手套
⑥鑷子
⑦衛生紙

①白色、茶色、深淺花棕色的柔軟毛皮布料
②白色長毛布料
③米白色貴賓絨毛布料

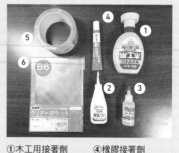

①木工用接著劑
②極細口噴嘴手工藝
　接著劑
③防綻液
④橡膠接著劑
⑤OPP膠帶
⑥OPP塑膠袋

製作步驟

# 描繪形象草稿、製作骨架

### Memo

整體身長輪廓設計成龍形，外表形象則塑形成實際存在且氣勢強大的野生動物。在製作虛構的生物時，會留意參考真實動物的身體構造。這次的「鹿龍」參考了馴鹿等鹿科動物，製作出有巨大犄角和有蹄的前腳。為了突顯幻獸的形象風格，參考猛禽類動物的身體，在龍的後腳跟添加了趾距，並且添加了只有冬天才會變成藍色的馴鹿眼睛。

製作作品前先描繪形象草稿。

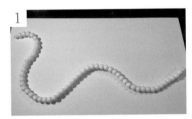

1 依照製作尺寸剪下所需的娃娃骨架。

2 剪下金屬線，用老虎鉗壓緊扭轉成較粗的鐵線。

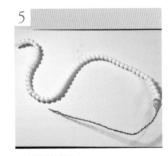

3 將步驟2製作的粗鐵線，纏繞在娃娃骨架的尾端後，扭轉成一條線。

4 將細金屬線捲繞在娃娃骨架和鐵線的連接處加以固定，並且用熱熔膠槍在表面塗滿熱熔膠，包覆住整個金屬線牢牢固定。

5 完成尾巴的部分。

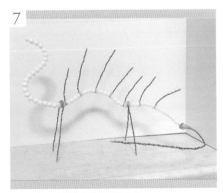

6 將較粗的鐵線對折扭轉後，捲繞在娃娃骨架上等同腳的位置，並且和尾巴一樣，用細金屬線和熱熔膠固定。在等同背棘的位置也重複這個步驟，將金屬線固定在此。

7 娃娃骨架和金屬線構成的骨架完成。

將手工藝棉花捲繞在骨架上，並且用羊毛氈戳針戳刺棉花加以固定。
※請注意戳針不要戳刺到娃娃骨架。

考量毛皮布料的厚度，一開始添加棉花時要塑形成比成品細窄的身形。

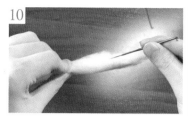

尾巴末端的金屬線也要捲繞上手工藝棉花，再輕輕戳刺固定後，用毛線綁緊。

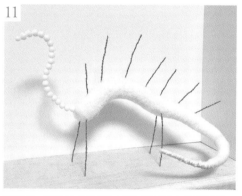

骨架完成。脖子要先製作出黏土部件後再捲繞上手工藝棉花。

# 眼睛製作

將石粉黏土按壓在半圓球模具做出造型。

用圓珠或半圓球玻璃的凸面按壓出圓弧狀，形成等同眼珠部分的凹陷形狀，再雕出虹膜和瞳孔。

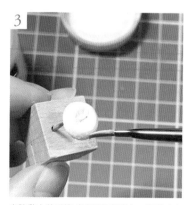

去除黏土的毛邊（不要的部分），讓黏土完全乾燥後再塗上打底劑。

打底劑乾了之後先從暗色顏料開始上色。整體塗滿壓克力漆（消光黑）後，依序塗上天空藍、青銅色、金色和銀色的壓克力顏料。

在瞳孔和虹膜的凹槽內倒入樹脂液，並且用牙籤去除氣泡，再用UV燈照射硬化。
※使用樹脂液時要穿戴手套。

完全硬化後再塗上樹脂液，使外觀從側面看時會呈圓拱狀並且使其硬化。最後包括眼白，整顆眼睛都塗上樹脂用表面修補液，並且使其硬化。

眼睛完成。

## 頭部製作

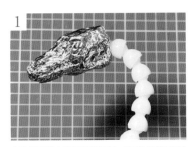

在娃娃骨架末端包裹鋁箔紙，做成頭部的芯材。

在鋁箔紙上包裹石粉黏土。在眼睛反面塗上薄薄的木工用接著劑後，配置在黏土上。

為了呈現不同質感，眼皮使用樹脂黏土（以1：1混合MODENA和GRACE）製作。用塑形刀雕塑形狀，等其稍微乾燥後，在眼皮以外的地方覆蓋上石粉黏土，做出眼睛周圍的造型。

用塑形刀、刮刀、湯匙狀刮刀，塑造出口鼻部、嘴巴周圍和脖子周圍的造型。細微的凹凸部分，建議以用水稍微沾濕的畫筆修飾表面。

**5**

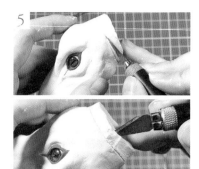

用工藝刀在脖子劃線挖出一道凹槽，當作之後黏貼毛皮的部位。

**6**

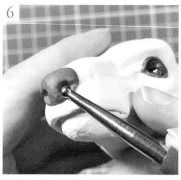

在黏接鼻子的位置塗上薄薄的木工用接著劑，再添加上樹脂黏土（以1：1混合MODENA和GRACE）後塑形，並且用前端為圓形的工具雕出鼻孔的形狀。

**7**

用塑形刀的紋路部分輕輕按壓鼻子，表現出皮膚的質感。

**8**

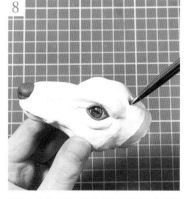

趁黏土未乾之際，用錐針在黏接鹿角的位置開孔。

**9**

黏土完全乾燥之後用砂紙、海綿砂紙，依照由大至小的號數打磨。

**10**

在黏土板上將黏土擀薄，再切割出耳朵的形狀。

**11**

在黏貼面塗上薄薄的木工用接著劑，將耳朵黏接在頭部。要將耳朵深埋黏接在根部溝槽以免乾燥後斷裂。

**12**

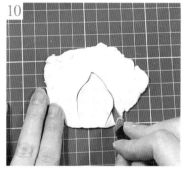

耳朵也要用砂紙打磨。細節部分用鑷子夾住砂紙碎片打磨。

**13**

塗上打底劑修飾表面。

# 鹿角製作

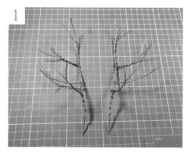

1

用金屬線製作鹿角的芯材。

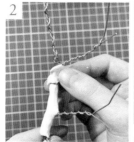 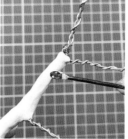 

2

用手指將石粉黏土壓扁包覆金屬線，並且用湯匙狀刮刀抹平修整。金屬線的末端也要完全覆蓋上黏土。

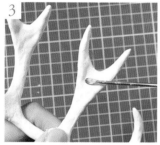 

3

鹿角分支的交界要塑造出明顯的凹狀，並且用湯匙狀刮刀刨出凹凸紋路。

4

黏土完全乾燥後用砂紙打磨鹿角末端，再以用水沾濕的畫筆修飾表面凹狀處。待完全乾燥後塗上打底劑修飾表面。

# 腳部製作

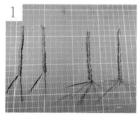

1

用金屬線製作腳的芯材。

2

先做出前腳。用石粉黏土包覆金屬線。

3

用樹脂黏土（以1：1混合MODENA和GRACE，並且加入茶色的壓克力顏料）製作出懸蹄，並且用木工用接著劑黏在腳跟處。接著在表面包覆一層石粉黏土。

4

在樹脂黏土（以1：1混合MODENA和GRACE，並且加入茶色的壓克力顏料）製成的蹄插入牙籤當成芯材，再用木工用接著劑接合在腳上。腳趾間隙也要填滿黏土，上部也要覆蓋石粉黏土。

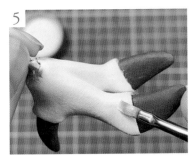

5

用砂紙打磨，再塗上打底劑修飾表面。

6

接著製作後腳。和前腳相同，用黏土包覆金屬線。

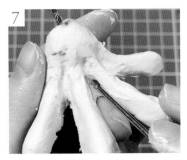

7

用湯匙狀刮刀刨出腳趾的間隙。

8

用塑形刀的紋理部分和揉皺的鋁箔紙，營造出皮膚質感。

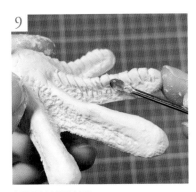

9

在腳趾刨出鱗狀紋路。

10

用樹脂黏土製作腳爪後與腳趾接合，再用黏土覆蓋上部。

11

腳底也要雕出腳趾的形狀，用塑形刀的紋理部分添加紋理。

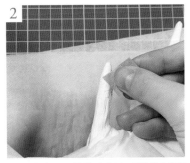

12

用濕的畫筆修飾表面，並塗上打底劑。

## 背棘製作

1

用黏土包覆金屬線芯材，並且在表面刨出凹凸紋路。

2

黏土乾燥之後用砂紙打磨末端，用沾濕的畫筆修飾凹凸紋路。※用衛生紙包覆本體（手工藝棉花包覆的部分）以免本體沾染粉塵。

3

塗上打底劑修飾表面。

## 頭部塗裝

慢慢在整體重疊塗上稍微稀釋後的壓克力顏料（在熟褐混入少量黑色）。

將步驟1的壓克力顏料用水稍微稀釋後為鼻子塗色。乾燥後先塗上半光澤水性壓克力保護漆，再塗上消光水性壓克力保護漆。

在步驟2的壓克力顏料添加少量的紅色（緋紅色、中國紅）為眼睛塗色，接著先塗上半光澤保護漆，再塗上消光保護漆。為了呈現水潤感在淚溝和眼睛內膜邊緣塗上光澤保護漆。

整顆頭塗上消光保護漆修飾。

## 鹿角塗裝

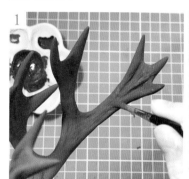

底色塗上和頭部相同的壓克力顏料後，再重疊塗上象牙色（灰米色＋白色）。

用乾的畫筆沾取壓克力顏料（在步驟1的顏色再添加白色）輕輕刷過凹凸紋路，為末端上色（乾刷）。越往末端越加重白色的用色形成漸層色調。

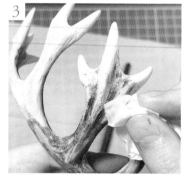

用水稀釋和頭部同色的壓克力顏料後，塗在鹿角表面，用沾濕的衛生紙輕拍表面擦拭，為鹿角紋理和凹凸間隙上色。

乾燥後先塗上半光澤保護漆，再塗上消光保護漆修飾。

## 腳部塗裝

1

先為前腳上色。腳爪以外的部分塗上和頭部顏色相同的壓克力顏料。用細畫筆盡量塗滿顏色，避免看到底層的白色黏土。

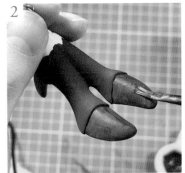

2

深茶色壓克力顏料加入黑色再用水稀釋後塗在整隻腳爪，並且用沾濕的衛生紙擦拭，營造出質感。

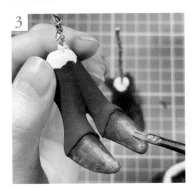

3

塗上米色（灰米色＋白色，再用水稀釋），表現出沾染泥土的樣子。

4

只有腳爪部分先塗上半光澤保護漆，再塗上消光保護漆修飾。※腳爪以外的部分和骨架相接後再塗上保護漆。

5

重複步驟1用相同的方法為後腳上色。

6

重複步驟2～4，用和前腳爪相同的方法為後腳爪上色。

7

在趾間和鱗狀表面重疊上色。

Pick up!

PADICO
水性壓克力保護漆sealer
厚塗光澤／半光澤／消光

這是用於塗裝修飾的保護漆。依照「半光澤」→「消光」的順序重疊塗上就可以呈現恰到好處的質感。

腳底、腳趾側邊、鱗狀間隙都塗上米色（灰米色＋白色，再用水稀釋）。

整體塗上半光澤保護漆，除了腳的鱗狀表面，其他部分都塗上消光保護漆修飾。

## 背棘和細節部分的塗裝

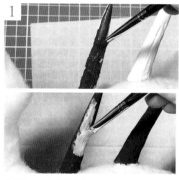

使用和P.62鹿角顏色相同的壓克力顏料，並且以相同方法上色。

用乾的畫筆沾取顏料，輕輕刷過凹凸表面，為末端上色（乾刷）。

用水稀釋和頭部顏色相同的壓克力顏料後，塗在表面，並且用沾濕的衛生紙輕輕拍擦拭上色。

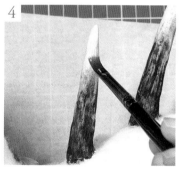
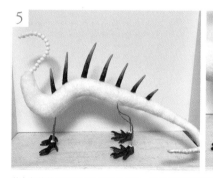

乾燥後先塗上半光澤保護漆，再塗上消光保護漆修飾。

將上色後的前腳和後腳安裝於骨架，再做出腳跟的造型並且上色。

完全乾燥後，用手工藝棉花包覆兩隻腳和脖子，整個玩偶的基底完成。頭部尚未黏接先拆除。

Pick up!

清原株式會社
手工藝棉花

纖維緊密交織的成團棉花，用於布偶填充或做出輪廓。除了用手撕下少量棉花使用，也可整張薄片（片狀棉花）使用，所以可當作鋪棉使用。

# 紙型製作～毛皮縫製

用保鮮膜包覆骨架後用膠帶黏貼固定，接著用筆描繪出身體每個部位的紙型。

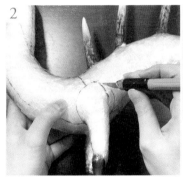

用工藝刀沿著線條裁切。為了做出左右對稱的樣子，只要裁切一邊的輪廓即可。

依照裁切的保鮮膜，描繪在紙板上做出紙型。變形或左右不對稱的部分，將紙對摺後稍微修剪。先在紙型標註毛皮布料的毛流。

在毛皮布料描繪紙型。

保留約5mm的縫份後，裁剪毛皮（從描繪的線條外側5mm處裁切）。這時一定要從反面裁切毛皮布料。

毛皮布料裁切後，在縫份塗上防綻液和薄薄的手工藝接著劑，並且等其完全乾燥。

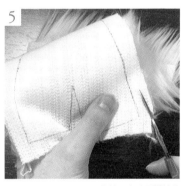

用全迴針縫縫合腳、身體、尾巴等各個部件。

裁切出身體部件的圖案形狀，並且塗上手工藝接著劑為邊緣做防綻處理。

依照裁切的圖案，準備相同形狀的白色毛皮布料。

10

用鎖邊縫將身體部件和白色毛皮布料縫合。

11

翻面後的樣子。

12

所有的部件。

13

用珠針固定脖子、身體、尾巴部件以免位移，並且用全迴針縫一一縫合。為了將毛皮布料穿上本體背後先不縫合。

14

縫合時用錐針將捲入的毛挑出表面。

15

將毛皮布料穿上本體。

16
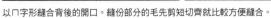

以ㄇ字形縫合背後的開口。縫份部分的毛先剪短切齊就比較方便縫合。

17

脖子周圍還要和頭部相接所以先留一個小開口。

18

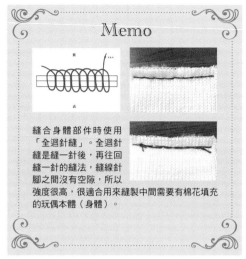

## Memo

縫合身體部件時使用「全迴針縫」。全迴針縫是縫一針後，再往回縫一針的縫法，縫線針腳之間沒有空隙，所以強度很高，很適合用來縫製中間需要有棉花填充的玩偶本體（身體）。

腳和尾巴部件以∏字形縫合。

## 修剪

用剪刀稍微修剪身體、尾巴、腳的毛。腳的部分，請剪出圓弧平順的毛流，才看得出膝蓋的形狀。

2

在腳的毛皮布料反面塗上橡膠接著劑，並且黏接在黏土部件上。

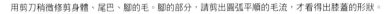

3

身體和腳的修剪完成。脖子周圍和頭部接合後再修剪。

## 將鹿角部件和娃娃骨架接合在頭部

1

用剪刀剪開脖子周圍尚未固定的手工藝棉花，並且暫時拆除和頭部相接的娃娃骨架。

2

用環氧樹脂將頭部和鹿角相接，內部的溝槽也要填補上環氧樹脂。

3

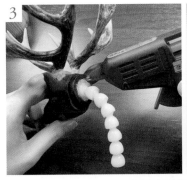 

待環氧樹脂完全乾燥後，用熱熔膠將頭部和拆除的娃娃骨架相接。

4

用砂紙打磨環氧樹脂的部分。

5

用和鹿角相同顏色的壓克力顏料為環氧樹脂的部分上色修飾。

## 植毛

1

 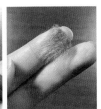

從毛皮布料的根部剪下毛，做成植毛用的塊狀纖維。

2

將剪下的毛鋪在OPP塑膠袋，在切口塗上手工藝接著劑。這時為了避免分散用牙籤塗上接著劑，連毛的反面都要完全塗上接著劑。待完全乾燥後再剪去多餘的接著劑。

3

塗上手工藝接著劑，用牙籤將塊狀纖維黏貼在頭部本體。植毛時直接在耳尖等本體細節處塗上接著劑，再用鑷子夾起毛後黏上。在這個步驟中很容易將纖維黏得到處都是，所以使用可擠出少量接著劑的極細口噴嘴接著劑。

4

前腳也用相同方法植毛。

5

整體植毛完成後，用剪刀稍微修剪。

## 將頭部和身體接合後，塗裝修飾

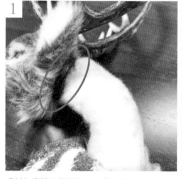

重新包裹脖子周圍的手工藝棉花，將頭部和本體的娃娃骨架相接，在頭部凹槽部分（上圖用紅色圈起的部分）塗上橡膠接著劑後黏上一圈毛。毛皮布料黏貼在脖子時的縫份大約 5 mm，比凹槽的寬度寬一些。

縫合脖子剩餘的部分，步驟 1 黏貼在頭部的毛皮布料也要和脖子縫合。

修剪脖子周圍的毛。

在毛皮的顏色交界上色。用畫筆塗上壓克力顏料（添加黑色的茶色），再用牙刷慢慢刷開。在大面積上色時，可以用針梳刷開上色。

在臉部周圍、前腳有黏貼毛的地方，用刷開的方式上色，毛會脫落，所以慢慢一點一點用畫筆上色抹開。上色過度的地方，用稍微沾濕的衛生紙輕輕擦拭調整。

用細畫筆在眼睛周圍塗上壓克力顏料（添加黑色和紅色的茶色）。

用針梳梳理毛流修整後即完成。

# 狼龍

2018年的羊毛氈作品。結合狼與龍兩種樣貌的幻獸，身體表面稍微植入一些羊毛，以全身長滿狼毛來取代龍鱗。底座以幻獸棲息在陡峭山嶽為意象，完全以手作製成。

# 小鳥幻獸 Nyx　中山和美

中山和美主要以泰迪熊玩偶的傳統技法，製作出創作人偶和有臉玩偶。
類型包括貓咪或兔子等動物到虛構幻獸，還有暗黑形象的作品，
不翻模、不量產，而是在關西的一隅默默創作出世上唯一的作品。

［HP］http://www.switchbladesister.com/
［Instagram］necoco_kazumin
［Twitter］@iiinecocoiii

# 小鳥幻獸Nyx的製作方法

道具

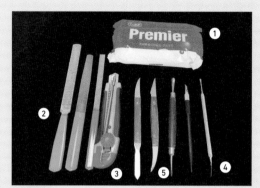

①石粉黏土
②金屬銼刀
③美工刀
④塑形刀
⑤黏土刮刀

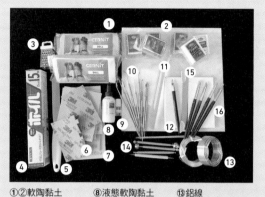

①②軟陶黏土
③磨泥器
④鋁箔紙
⑤刮刀
⑥海綿砂紙
⑦砂布
⑧液態軟陶黏土
　（液態樹脂黏土）
⑨樹脂黏土軟化劑
⑩塑形刀
⑪壓克力擀棍
⑫工藝刀
⑬鋁線
⑭球狀黏土工具
⑮黏土刀
⑯矽膠型黏土刮刀

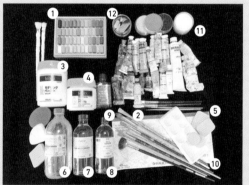

①粉彩顏料
②畫筆
③塑型劑
④凝膠劑
⑤海綿
⑥調和油
⑦⑧⑨油畫顏料畫
用油（松節油）
⑩調色盤
⑪油畫顏料
⑫碎海綿

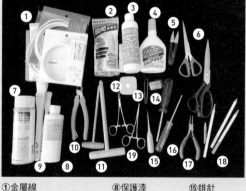

①金屬線
②瞬間接著劑
③模型用膠狀接著劑
④速乾型多用途接著劑
⑤紗線剪
⑥剪刀
⑦光澤噴霧
⑧保護漆
⑨刷子
⑩斜口鉗
⑪塞棉推
⑫量尺
⑬鑷子
⑭毛海刷毛器
⑮錐針
⑯螺絲起子
⑰老虎鉗
⑱粉土筆
⑲返裡鉗

材料

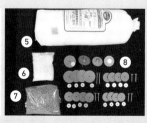

①手縫線
②毛海布料
③羽毛
④羽毛緞帶
⑤手工藝棉花
⑥木屑顆粒
⑦不鏽鋼珠
⑧關節套組
⑨玻璃眼珠

製作步驟
# 臉部製作（塑形）

用力搓揉石粉黏土，用擀棍擀平後裁切成橢圓形。

手指用水沾濕，推開延展黏土。像在製作有邊緣的盤子，將邊緣的黏土往中間推，使中央從表面看宛如微微突出的圓拱狀。

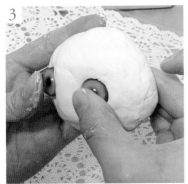

將玻璃眼珠按壓在眼睛的位置。

切去多餘的黏土，調整臉部線條。

額頭、鼻梁、眼睛周圍補上黏土。

手指用水沾濕後將黏土推開相接合，並且用塑形刀修整。完成大概的造型後，先讓其乾燥。

製作眼睛周圍的造型。

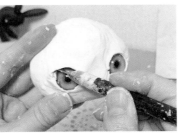

製作鳥喙狀嘴型。額頭～眼睛上方的黏土分量有點不足，所以再添加一些黏土。用手指推勻添加的黏土後，用工藝刀或塑形刀塑造出唇線。之後修飾整體時，會再做出細節的形狀，所以這裡只要做出大概的形狀即可。

為了讓臉部從側面看時，鳥喙狀嘴巴更往前突出，所以又添加一些黏土，並且用手指捏成尖狀。

用金屬銼刀將脖子後面磨平，變成半圓形。

整體先用砂布打磨後，再用更細的海綿砂紙打磨。

臉部造型完成。

Pick up!

PADICO
Premier高級輕量石粉黏土

這是一種質輕、堅固又容易塑形的黏土。小型作品有時會使用La Doll石塑黏土，不過大型作品的臉部最好使用較輕的黏土所以建議使用Premier高級輕量石粉黏土。這款黏土乾燥後依舊保有強度，所以我認為是石粉黏土中最適合用於製作藝術玩偶的選擇。

## 臉部製作（上色）

用較柔軟的畫筆，在眼睛內側和整張臉塗上液狀打底劑（混合塑型劑和凝膠劑）。塗1～2次後讓其乾燥。

用海綿輕輕拍打上膠狀打底劑（將塑型劑和凝膠劑混合後，打發成如鮮奶油般的硬度）

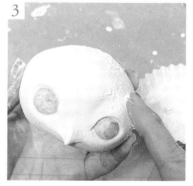

用手指點塗營造出紋理。多次重疊塗上後，讓其乾燥。

用海綿砂紙打磨鳥喙狀嘴巴，去除粗糙感。

混合油畫顏料的畫用油和油畫顏料後，塗滿整張臉，再讓其乾燥。

6

一邊想像完成的樣子，一邊用調和油和油畫顏料塗上大致的顏色。
眼皮內側塗較深的顏色，眼睛周圍塗上粉紅色。整體上色完成後，先讓其乾燥。

7

 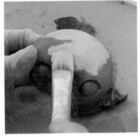  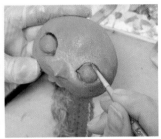

參考選用的毛海布料色彩，並且調配出顏色，利用畫筆、手指、海綿重疊上色。不要只塗一種顏色，而是一邊混合各種顏色一邊上色。
為眼皮內側上色時請小心不要超過邊界。整體上色完成後讓其乾燥。乾燥後為鼻子和唇線等細節上色後，再次讓其乾燥。

8

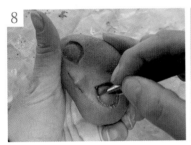 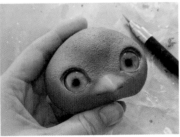

玻璃眼珠的部分用水沾濕軟化後，用工藝刀在邊緣劃出切口，剝除覆蓋在玻璃眼珠的塗裝。

9

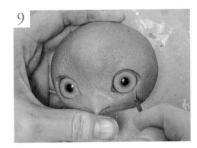

為眼皮內側上色。

10

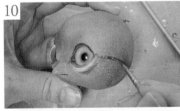

添加圖案。

11

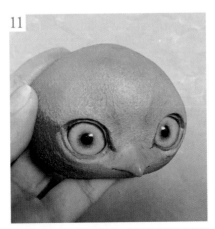

調整整體色調。添加眼線後，靜置到顏料完全乾燥
並且不要觸碰，完全乾燥後即完成。

### Memo

關於乾燥時間，夏天最好放
置約 1 週以上，冬天則須放
置約 2 週的時間。
要備齊所有的油畫用品有點
困難，所以初期製作時，也
可以使用壓克力顏料來修飾
上色。

# 紙型製作～毛海剪裁和縫製

1

依照臉部尺寸描繪出紙型設計。這時就要先決定連接部分使用的關節尺寸。

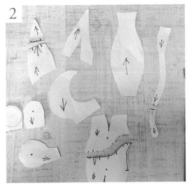

2

標記毛海的毛流方向後，裁切紙型。

3

將紙型描繪在毛海布料後裁切下來。因為紙型沒有包含縫份，所以在紙型輪廓外側保留約 5 mm縫份後裁切。毛海的毛流較長，所以請小心不要剪到毛。

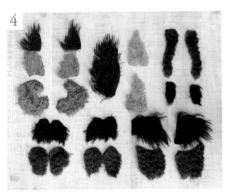

4

所有部件裁切好的樣子。

5

有拼接的部件，先縫合拼接的部分。

6

身體正面相對用半迴針縫縫合。

7

頭部縫合打摺處後，將中央部分和左右部件縫合。

8

縫合後翻回正面。尾巴等細小部件用返裡鉗夾住內側後拉出翻回。

## Memo

毛海是用安哥拉山羊的毛製成的布料，特色是擁有生絲般的光澤，毛流長、透氣性佳、纖維的彈性佳。雖然是價位頗高的布料，但是堅韌、不易掉毛、不易沾染汙漬、不易褪色等，是相當耐用的絕佳天然材料，因此從以前就會用於製作泰迪熊玩偶。

# 手臂製作

取適量的羽毛裝飾排列成漂亮的樣子。

用模型用膠狀接著黏合上部。

用顏色不顯眼的縫線縫固定。

將裁切好的布料正面相對，以半迴針縫縫合後，翻回正面。這時保留安裝關節套組的開口和添加羽毛部件的地方，先不要縫合。

用錐針在安裝關節套組的部分開孔，裝上關節套組。

從下側的開口放入步驟 3 做好的羽毛部件。

一邊將毛海布料的縫份往內摺，一邊將羽毛部件和布料縫合。

以ㄇ字形縫法縫合安裝關節套組的開口。

手臂部件完成。

# 腳部製作

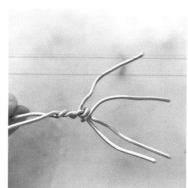
依照設計圖剪下鋁線，用 3 條鋁線捲繞成腳的芯材。

80

**2**

將 2 條較長的鋁線往上捲繞在一起,末端彎折捲起。腳爪部分稍微彎曲。至此芯材完成。

**3**

用力搓揉軟陶黏土。這次混合「Pardo珠寶黏土」的黑曜岩和半透明兩種顏色做成腳爪,使用縞瑪瑙做成腳。將軟陶黏土揉至耳垂的硬度後,先搓成球狀。

**4**

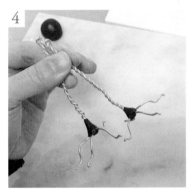

將黏土包覆在鋁線容易鬆動的地方,如照片般固定。

**5**

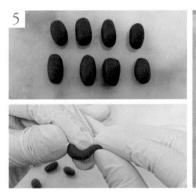

準備 8 個混合黑曜岩和半透明 2 種顏色的軟陶黏土,做成腳爪的形狀。

**6**

將做好的腳爪穿進鋁線的末端接合。

**7**

再以120度預熱的旋風烤箱內燒製約25分鐘。

8

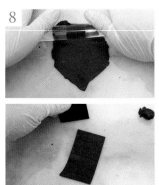

將縞瑪瑙軟陶黏土擀薄後分等，並且包覆腳趾。將黏土從下側包覆住鋁線，上側也添加黏土遮覆住交界線。整隻腳用 2 片黏土包覆做出腳形。

9

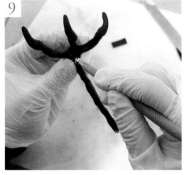

用手指或球狀黏土工具抹平黏土的交界，使表面平整。

10

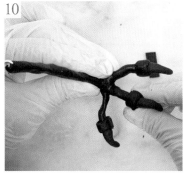

將黏土捲覆在腳爪和腳趾之間，用球狀黏土工具平整表面。

11

用黏土包覆腳踝添加分量。

12

用黏土刮刀添加紋理。這個步驟完成之後，再用相同方法製作另一隻腳。

13

用碎海綿為趾尖塗上粉彩顏料。

14

一邊確認整體，一邊添加細節紋理。

15

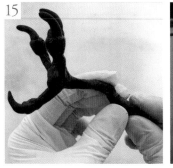

調整腳的角度後，放入以120度預熱的旋風烤箱中燒製40分鐘。

16

腳的軟陶黏土部分即完成。

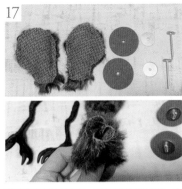

**17**

裁好的布料正面相對，以半迴針縫縫合後，翻回正面。這時保留安裝關節套組的開口，以及步驟16接合腳的部分先不縫合。

**18**

用錐針在安裝關節套組的部分開孔，裝上關節套組。

**19**

從下側的開口放入步驟16的腳。

**20**

在軟陶黏土製的腳上部塗上瞬間接著劑，一邊將毛海布料往內摺，一邊黏合。

**21**

從開口塞入手工藝棉花。

**22**

以ㄇ字形縫法縫合。

**23**

用毛海刷毛器梳理毛，讓接縫處變得不顯眼。完成另一邊的部件，並且留意左右對稱，腳的部件即完成。

Pick up!

Pardo珠寶黏土
黑曜岩／半透明／縞瑪瑙
提供：株式會社SUN-K

這是一種有加入蜜蠟的軟陶黏土，色彩多樣方便好用。特色是燒製後會稍微變色，我自己個人使用的感想是，燒製時不會失敗，製作時也好操作，所以是我最常用的軟陶黏土。很適合用來製作小型部件。

## 尾巴製作

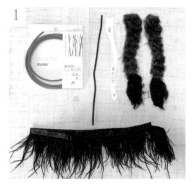

1 準備比尾巴稍長的金屬線（自遊自在）。

2 羽毛緞帶塗上瞬間接著劑後捲起，並且縫合固定在金屬線末端。

3 在毛海布料只有縫合拼接部分和一邊輪廓線的狀態下，順著整條金屬線捲繞縫合。

4 開口部分用縮口縫收緊縫合。

5 尾巴完成。

## 耳朵製作

1

縫合耳朵內側的打摺。

2 將毛海布料的耳朵部件正面相對後以半迴針縫縫合。

3 將金屬線（自遊自在）依照耳朵的輪廓彎折後剪下。將兩端切口彎折捲起。

**4**

將縫合後的耳朵部件翻回正面。用錐針挑出接縫處夾住的毛。

**5**

將步驟 3 彎折的金屬線順著耳朵的輪廓塞入。

**6**

將耳尖的毛沾濕扭轉成束。

**7**

縫合開口。這時為了避免金屬線位移,將金屬線縫合在毛海布料上。

**8**

耳朵完成。

**9**

至此所有部件皆備齊。

Pick up!

日本化線
自遊自在

可簡單彎折成自己喜歡的形狀,又很強韌,所以我很喜歡使用。顏色和粗細的種類豐富,具耐久性卻能用手彎折成想要的形狀,所以是相當好用的金屬線。

## 部件組合和最後修飾

**1**

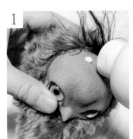 

在頭部塗上模型用膠狀接著劑,一邊將毛海布料的裁切面往內摺一邊覆蓋在頭部。下顎同樣塗上接著劑後,和毛海部件黏合。

**2**

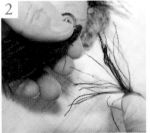 

將黑色長毛海收成一束後沾取接著劑,再用錐針塞入頭部正中央。

3

從開口塞入手工藝棉花。

4

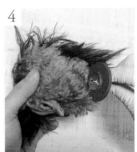 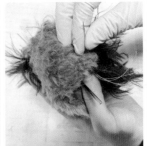 

將關節套組安裝在脖子，再用平針縫將開口縫合收緊。

5

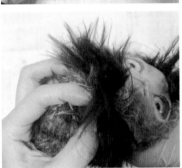

在身體關節位置的中心，用錐針開孔，從身體外側插入T針。

6

 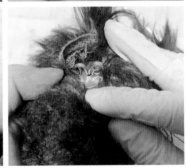

從身體內側依序放入硬質紙盤→鋼盤，再用螺絲起子將關節針（T針）往上捲，一直捲緊到底部，使關節套組不會鬆脫。

7

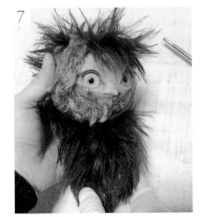

將頭部和身體結合。

8

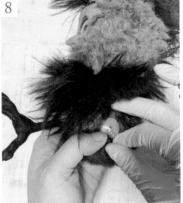 

接著重複步驟5～7接合手腳。

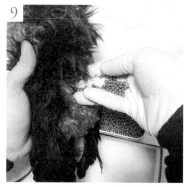

9

從背後開口放入不鏽鋼珠、手工藝棉花、木屑顆粒。填充時將不鏽鋼珠往身體下方,木屑顆粒和手工藝棉花往身體上方塞入。

10

填充後,以ㄇ字形縫法縫合背後。

11

考量整體比例決定耳朵接合的位置,並且用珠針固定縫合。

12

用毛海刷毛器梳理,讓接縫處變得不顯眼。

13

調整頭部中央的黑色長毛海並且立起。

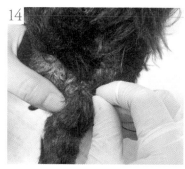
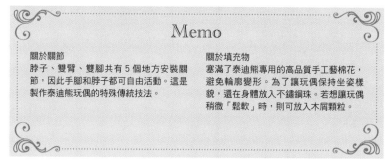

14

縫上尾巴即完成。

## Memo

關於關節
脖子、雙臂、雙腳共有5個地方安裝關節,因此手腳和脖子都可自由活動。這是製作泰迪熊玩偶的特殊傳統技法。

關於填充物
塞滿了泰迪熊專用的高品質手工藝棉花,避免輪廓變形。為了讓玩偶保持坐姿樣貌,還在身體放入不鏽鋼珠。若想讓玩偶稍微「鬆軟」時,則可放入木屑顆粒。

# 展翅幻獸Sunny 中山和美

工具

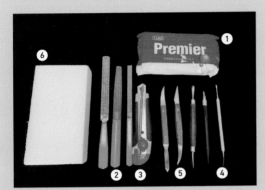

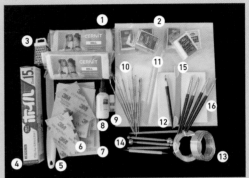

①石粉黏土
②金屬銼刀
③美工刀
④塑形刀
⑤黏土刮刀
⑥保麗龍

①②軟陶黏土
③磨泥器
④鋁箔紙
⑤刮刀
⑥海綿砂紙
⑦砂布
⑧液態軟陶黏土（液態樹脂黏土）
⑨樹脂黏土軟化劑
⑩塑形刀
⑪壓克力擀棍
⑫工藝刀
⑬鋁線
⑭球狀黏土工具
⑮黏土刀
⑯矽膠型黏土刮刀

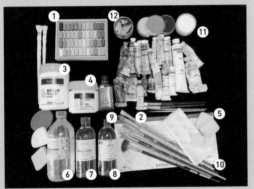

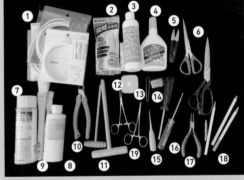

①粉彩顏料
②畫筆
③塑型劑
④凝膠劑
⑤海綿
⑥調和油
⑦⑧⑨油畫顏料的畫用油（松節油）
⑩調色盤
⑪油畫顏料
⑫碎海綿

①金屬線
②瞬間接著劑
③模型用膠狀接著劑
④速乾型多用途接著劑
⑤紗線剪
⑥剪刀
⑦光澤噴霧
⑧保護漆
⑨刷子
⑩斜口鉗
⑪塞棉推
⑫量尺
⑬鑷子
⑭毛海刷毛器
⑮錐針
⑯螺絲起子
⑰老虎鉗
⑱粉土筆
⑲返裡鉗

材料

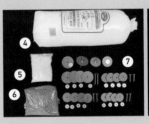

①手縫線
②毛海布料
③羽毛
④手工藝棉花
⑤木屑顆粒
⑥不鏽鋼珠
⑦關節套組
⑧玻璃眼珠

# 臉部製作（塑形）

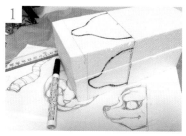

製作臉部口鼻部較長的幻獸時，使用保麗龍製作芯材。依照設計圖，在保麗龍上畫出輔助線。

用美工刀削切出大致的形狀。邊角和曲線部分用金屬銼刀前端削磨。眼睛周圍用銼刀尖端挖出輪廓。

用保鮮膜包裹保麗龍芯材。

從袋中取出石粉黏土，用手使勁搓揉。利用身體的重量，用手掌按壓直到表面平滑。

用擀棍擀平推開至0.5～1cm左右的厚度。

用黏土包覆保麗龍芯材。用手指按壓內凹處，讓黏土緊密服貼在保麗龍表面，呈現明顯的凹凸輪廓。

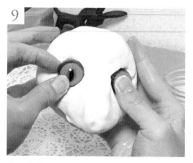

切除多餘的黏土。

手指沾水，用手指和刮刀修飾整體，使表面平滑。

塞入玻璃眼珠。
※如果使用壓克力或矽膠製的眼珠，剝除打底劑時較不容易剝離，所以避免使用。

劃出大概的唇線輪廓。

在額頭、眼睛的上下側添加黏土，用沾水的手指或刮刀撫平接合。

用塑形刀削切出口鼻部的輪廓，並且做出鼻尖的造型。

在眼睛周圍和臉頰一點一點補上黏土，用手指抹平推開。若添加過多則削除。重複這些動作直到成型。

完成大概的形狀後使其乾燥。

大約需乾燥一天，在半乾的狀態下切除脖子部分的黏土。

完全乾燥後，再補上黏土調整形狀。這次覺得眼尾上部、額頭、臉頰下側、口鼻部的線條、鼻子下方隆起處皆不足，所以在這些地方添加黏土，並且用手指撫平，呈現自然的隆起輪廓。

調整時從上方確認口鼻部和左右臉頰是否大概同高，還有從下方確認嘴巴線條和眼皮突出處是否平均。調整完成後使其乾燥。

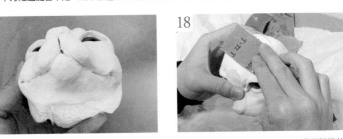

乾燥後用砂布打磨。輪廓還不夠清晰細緻的臉部線條和臉頰，依照100號→240號→400號的順序打磨。

若有刮痕或凹痕，用黏土修補。

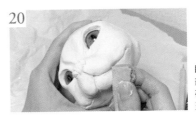

眼皮上方、口鼻部和唇線周圍用砂紙400號打磨至表面平滑。

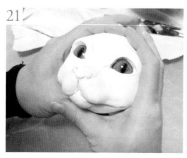
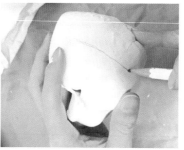
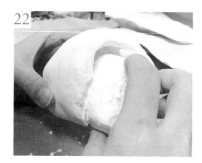

決定臉露出的面積後，在大約一根手指寬的位置標註記號。
※黏土和毛海的黏貼面約有1cm即可，所以1根手指寬即可。

拔出保麗龍。

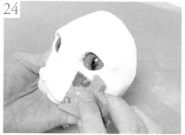

用美工刀切除多餘黏土。

用海綿砂紙打磨細節。一邊注意不要傷到玻璃眼珠，一邊修飾眼皮內側等細節部分，避免出現凹凸不平的表面。

## 臉部製作（上色）

在塑型劑中添加少量的凝膠劑，並且加水溶解後，當成打底劑使用。用柔軟的畫筆薄薄塗上。

盡量不要留下筆痕，輕輕仔細地塗上。眼皮內側、鼻子凹槽等細節處也不要遺漏，塗好後使其乾燥。

乾燥後再塗一次打底劑。這個步驟大概重複3次左右，重疊塗上打底。

第3次打底完成後，在完全乾燥之前，用不尖銳的塑形刀，沿著嘴巴和鼻子凹槽按壓出形狀。

完全乾燥後用海綿砂紙打磨。

6

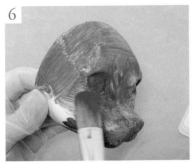 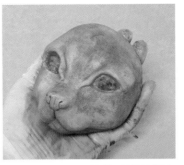

混合油畫顏料的畫用油和油畫顏料後，塗滿整張臉。用畫筆塗色後用海綿輕拍。這個步驟是在塗上底色，若有地方未上色，使顏色不均也沒關係。

7

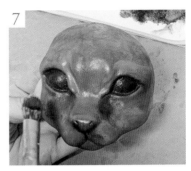

底色乾了之後，用調和油和油畫顏料塗上大概的顏色。

8

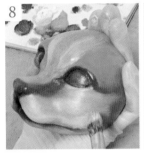 

眼皮內側、鼻子凹槽塗較濃的顏色。這個階段塗色時超過邊界也沒關係，大概塗色即可。其他部分也分別依照成品塗上大概的顏色即可。比起色彩的完成度，更注重明暗差異。眼睛邊緣和鼻尖塗上粉紅色後，使其乾燥。

9

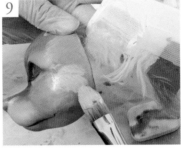 

參考選用的毛海布料色彩準備顏色，利用畫筆、手指、海綿重疊上色。不要只塗一種顏色，而是一邊混合各種顏色，一邊上色。為眼皮內側上色時，請小心不要超過邊界。整體上色完成後，讓其乾燥。

10

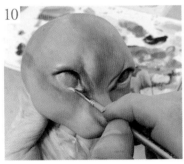  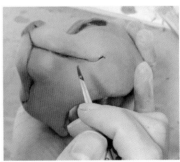

修飾眼線、鼻子、毛流和嘴巴等細節。

11

乾燥後，玻璃眼珠的部分用水沾濕。用筆刀在邊緣劃出切口，剝除覆蓋在玻璃眼珠的塗裝。

12

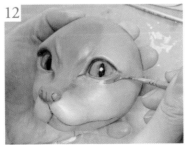 

在眼睛周圍補上顏色後即完成。

## 紙型製作～毛海剪裁和縫製

**1**

依照臉部尺寸描繪出紙型設計，製作各部位的紙型。這時也要先決定關節尺寸。
※翅膀和耳朵等整體完成後再調整尺寸，所以這裡先不做。

**2**

因為身體有厚度，所以先將做好的紙型用訂書針或膠帶接合做成立體造型，調整後完成紙型製作。

**3**

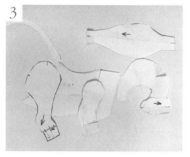

紙型也要標記毛海的毛流方向。

**4**

將紙型描繪在毛海布料後裁切下來。因紙型沒有包含縫份，所以在紙型輪廓外側保留約5mm縫份後裁切。毛海的毛流較長，所以請小心不要剪到毛。

**5**

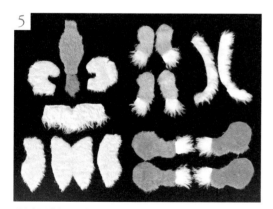

所有部件裁切好的樣子。因為毛海的毛流較長，為了方便縫合，先將縫份部分的毛剪短。

**6**

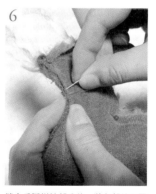

縫合手腳拼接部分後，將各部件正面相對，用珠針固定，再用半迴針縫縫合。

**7**

縫合後，在縫份剪出牙口。

**8**

將縫好的部件都翻至正面。細小部件用返裡鉗夾住內側後拉出翻回。

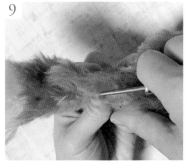  

所有部件翻至正面後，挑出接縫處夾住的毛並且梳順。毛流較長的部件用錐針慢慢挑出。毛流較短的部件則用毛海刷毛器梳理。

所有部件備齊的樣子。

## 犄角的製作

用鋁箔紙製作芯材。扭轉，捏緊做出形狀。

分割出所需的軟陶黏土分量，用力搓揉至黏土變軟。這次使用Pardo珠寶黏土的紫丁香蛋白石、紫水晶、透明色、半透明紅。先從紫丁香蛋白石黏土開始依序搓揉。

  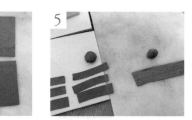

將黏土揉整成團不散落後，用壓克力擀棍擀開。折疊後再次擀開，如此多次反覆作業，直到黏土變成像耳垂般柔軟。

將黏土擀成平均一致的薄度後，裁切成長方形。

將紫丁香蛋白石和紫水晶混合後，以相同方法用力搓揉，再擀薄分割。

  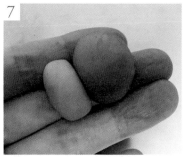

混合透明色和半透明紅。

將揉好的黏土和混合好的紫水晶黏合，做出漸層色調後，再擀薄分割。

97

8

將漸層黏土捲繞在鋁箔紙的芯材尖端包覆黏貼。依照紫水晶→紫水晶和紫丁香蛋白石混色→紫丁香蛋白石的順序捲繞分割好的黏土。

9

一邊捲繞，一邊在各處劃出橫線，表現犄角的質感。

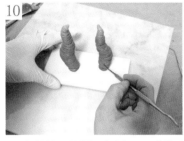 

10

用手拿著犄角，質感紋理會消失，所以若將近完成之際，不用手拿而是將犄角立在磁磚上繼續完成。在圓形黏土上塗少許液態軟陶黏土後，放在磁磚上，將犄角立起。這樣作業時就不用擔心犄角傾倒。

11

再以120度預熱的旋風烤箱中燒製50分鐘。
※燒製時間視軟陶黏土和成品大小、厚度而有不同，所以一邊觀察，一邊燒製。

12

只有犄角尖端沾水，再用防水的海綿砂紙（2000～5000號）輕輕打磨後即完成。

## 手腳製作

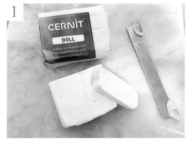 

1

先揉製所需的軟陶黏土。這次使用CERNIT DOLL白色、Pardo珠寶黏土的安地斯蛋白石，以及紫水晶和紫丁香蛋白石混合後的黏土。

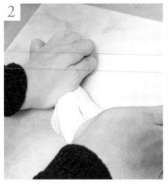 

2

用力擠壓搓揉CERNIT黏土，將空氣擠壓出去。折疊後再次擀開，如此多次反覆作業，直到黏土如耳垂般柔軟。搓揉好的黏土先依照所需的大概尺寸搓揉成球狀。

Pick up!

CERNIT DOLL白色
提供：株式會社SUN-K

這是一種有透明感的軟陶黏土。質地偏軟，或許對有些人來說不太容易操作，但是成品的質地光滑，適合用於表現人的肌膚。我會用於人型臉部的製作或希望呈現透明感的部件。

3

先製作腳的部件。分割搓揉好的CERNIT DOLL 白色黏土。建議在紙上描繪出大概的大小，一邊對照尺寸，一邊製作。

4

分割後的黏土用指尖搓揉成棒狀。

5

左右各準備4條，再將4條黏土黏合，一邊抹去接合處，一邊做出腳的形狀。

6

一邊添加腳踝和腳背的黏土，一邊捏出造型。

7

先做出接合腳底肉球的凹槽。

8

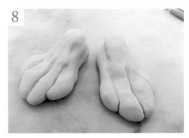

完成一隻腳後，做出另一隻成對的腳。

9

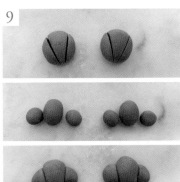

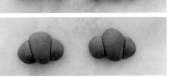

使用Pardo珠寶黏土的安地斯蛋白石做成肉球。將球狀黏土黏合後，用工具使接合處的表面平滑並且做出造型。

10

將球狀黏土壓扁，做出腳趾部分的肉球。

11

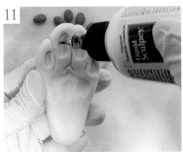

在步驟7做出的凹槽塗上薄薄的液態軟陶黏土，將肉球部件按壓黏上。

12

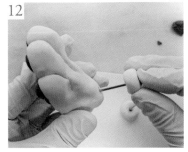

製作腳和本體接合時塗上接著劑的部分。將球狀黏土穿進剪短的鋁線部件後插入腳中。

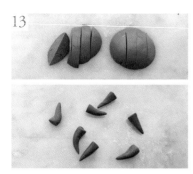

**13**

使用製作犄角時殘留的黏土（紫水晶和紫丁香蛋白石的混色）製作腳爪。用手指搓揉成細條狀，並且將末端彎曲成型。

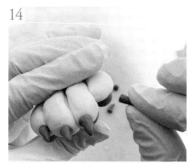

**14**

塗上少量的液態軟陶黏土，黏接在趾尖。

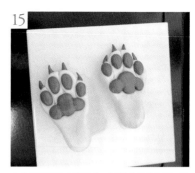

**15**

左右腳都黏接上腳爪後，調整整體的比例，放進120度的烤箱燒製約30分鐘。

**16**

取出腳爪並且冷卻後，用粉彩顏料塗色添加陰影。這次在趾尖塗上紫色系色調，整體塗上淡淡的橘色形成漸層色調。

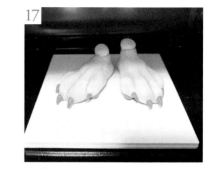

**17**

再次放入120度的烤箱燒製約30分鐘。

**18**

重複步驟３～17製作出手部。

**19**

在步驟12製作的黏貼面塗上瞬間接著劑覆蓋上毛海部件。用錐針將黏貼面的毛海布料往內摺。所有的手腳都用相同方法黏貼上毛海部件。

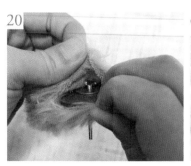

**20**

將手工藝棉花撕成小塊後，緊緊塞入手部前端約一半的位置。用錐針在關節安裝位置的正中央開孔後，安裝上關節套組再塞入棉花。

**21**

將手工藝棉花塞至開口處，用凵字形縫法縫合開口處。

22

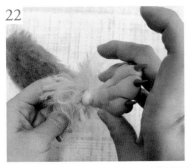

腳也重複步驟19～20黏貼毛海部件、安裝關節套組,並且塞入手工藝棉花。

23

24

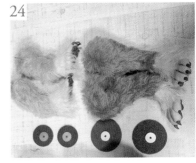

在腳靠近坐姿下側的部位放入不鏽鋼珠。先將不鏽鋼珠放入柔軟有彈性的袋中,再塞入毛海部件中。中間的空隙用化纖棉花塞滿固定。

手腳部件都備齊的樣子。

# 尾巴製作

1

2

3

將毛海收整成束。

準備3種各15～20cm的羽毛緞帶。

用羽毛緞帶捲繞成束的毛海並且縫合固定。不須太在意縫合的方法,只要避免成束的毛海和緞帶分離即可。

4

5

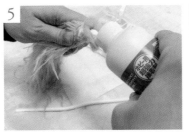

準備尾巴長度＋6cm左右的金屬線。

在毛海部件倒入模型用膠狀接著劑後將金屬線用力塞入。

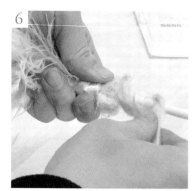

6

將多餘的毛海捲繞在金屬線後，用柔軟的橡皮繩捲繞，固定住金屬線和尾端部件。

7

條狀毛海部件縫合後翻回正面，再穿入整根金屬線，將尾端部件的根部和毛海部件的開口縫合。

8

從開口塞入手工藝棉花後，用ㄇ字形縫法縫合使接縫處不明顯。在尾巴接合根部有金屬線露出的狀態下縫合開口。

### Memo

尾端的毛海是將進口的獸毛洗淨後，用英國DYLON染料（PREMIUM DYE）染色。相當費工，所以若想簡單製作，建議可以使用假髮用的成束毛海。

---

## 將頭部、身體、手腳和尾巴接合

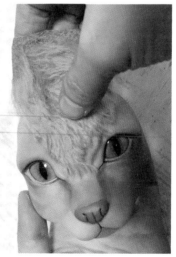

1

先在口鼻部塗上模型用膠狀接著劑後，黏貼上毛海布料。

2

在臉部周圍（上半部）塗上模型用膠狀接著劑，將毛海布料往內摺後黏貼。這時請注意不要將毛的末梢捲入。下半部從間隙塗上瞬間接著劑，按壓黏貼上布料。

3

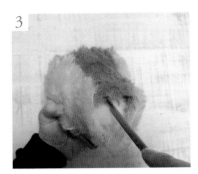

在頭部塞入手工藝棉花。

4

將關節套組安裝在脖子，再用平針縫將開口縫合收緊。打結後，為了避免關節鬆動周圍也縫緊固定，頭部部件即完成。

5

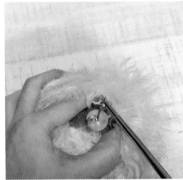

從身體外側插入T針，從身體內側依序放入硬質紙盤→鋼盤，再用螺絲起子將關節針（T針）往下捲。一直捲緊到底部，使關節套組不會鬆脫。

6

至此完成頭部和身體的接合。

7

腳和手臂也用相同方法和身體連接。

8

將不鏽鋼珠放入柔軟有彈性的布袋後，連同手工藝棉花塞入身體。

9

將放入不鏽鋼珠的布袋放在臀部，上半身則塞入手工藝棉花。重點是要塞滿至脖子周圍。

10

插入尾巴部件，從正面縫合固定。

11

確認玩偶是否可以穩定坐下後，用∩字形縫法
縫合背後。

12

用剪刀在頭部犄角的安裝位置剪出切口（切口
比犄角根部的尺寸小2倍左右），並且插入沾
有瞬間接著劑的犄角。

# 耳朵製作

1

確認整體比例決定耳朵大小。

2

做出外耳和內耳的紙型。分別描繪出2片左右對稱的耳朵後裁切布料。

3

縫合內耳的打摺後，將外耳和內耳的毛海布料
正面相對縫合。

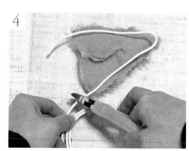

4

將金屬線依照耳朵的輪廓彎折後剪下。末端彎折捲起。

5

將耳朵翻回正面後，在耳尖縫上縫線後放入
金屬線。將金屬線縫合在耳尖的位置，避免
位移。

106

**6**

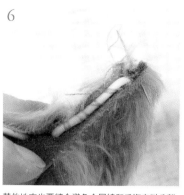 

其他地方也要縫合避免金屬線和毛海布料分離。開口縫合後耳朵部件即完成。

**7**

將耳朵和頭部接合。決定好接合位置後用珠針固定縫合。

**8**

另一隻耳朵也用相同方法縫合。將縫線收緊使接縫處不顯眼。至此兩隻耳朵都已縫合固定。

## 翅膀製作

**1**

依照設計圖製作紙型。粉紅色輪廓為毛海部分。

**2**

 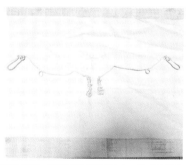

彎曲金屬線做成翅膀的芯材。建議使用堅韌耐用的金屬線。這次使用「頑固自在」和「自遊自在」的金屬線。一邊做好後再做出成對的另一邊芯材。

**3**

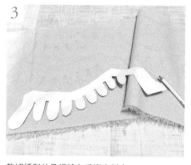

裁切紙型並且描繪在毛海布料上。

**4**

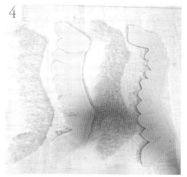 

裁切毛海布料後，正面相對縫合翅膀上部。

5

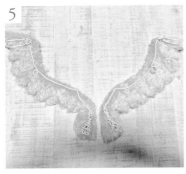

翻回正面，將金屬線沿著上部縫合固定避免毛海布料和金屬線分離。

6

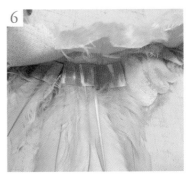

在毛海內側塗上速乾型多用途接著劑，夾入重疊的羽毛。

7

8

做出另一隻左右對稱的翅膀。

遮覆住毛海後，為了避免羽毛鬆脫以不明顯的縫法將羽毛和毛海縫合固定。毛海切口明顯的地方，連同縫份一起縫合。根部也要縫合收緊。

9

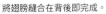

將翅膀縫合在背後即完成。

# 三尾狐 anya  @anyamals

這位個人創作者自幼就喜歡擬真布偶，2020年開始製作藝術玩偶。
以狐狸系布偶為主，製作並販售逼真又可愛的幻獸。

[Instagram] _anyamals
[Twitter] @macaronoxou

# 三尾狐的製作方法

工具和材料

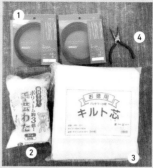
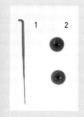

①輕量黏土
②③樹脂黏土
④塑形刀

⑤底座（保麗龍球和鋁箔紙）

①金屬線
（線徑2.6mm、
線徑1.6mm）

②手工藝棉花
③鋪棉
④老虎鉗

①戳針
②玻璃眼珠

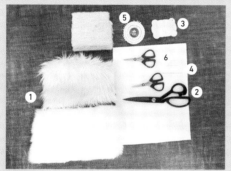

①毛皮布料　　④縫針
②裁縫剪刀　　⑤珠針
③手縫線　　　⑥剪刀

①水性壓克力保護漆
②粉彩定畫液噴霧

③速乾型多用途接著劑
④布偶鬍鬚部件

①粉彩顏料
②布用塗料
③畫筆
④棉花棒
⑤牙刷
⑥針梳

①噴筆

製作步驟 ·············································

# 臉部製作

1
用鋁箔紙包裹保麗龍球當成基底。

2
用樹脂黏土（白色）包覆基底製作臉部。做出大概的形狀後，在眼睛的位置埋入玻璃眼珠。

3
製作突出的口鼻部，用塑形刀塑造出鼻子和嘴巴的形狀。

4
調整好鼻子和嘴巴的形狀後，接著做眼睛。製作眼皮、眼睛邊緣、眼頭、眼尾凹陷處，做出眼睛的神韻。

5
大概放置半天到一天的時間乾燥。完全乾燥後將基底拆除。

6
用剪刀修剪多餘的黏土。

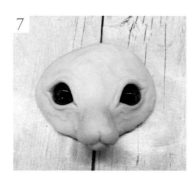

7
臉部的造型完成。

8

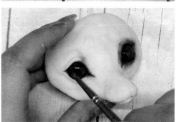

Pick up!

PADICO
MODENA

這是一種具有透明感和透光性的樹脂黏土。乾燥後堅固耐彎折，成品質感相當光滑。

接著開始塗色。因為臉上會貼上毛皮，所以幾乎會遮住黏土的部分，但是會從毛皮的切口縫隙看到少許的黏土，所以先用粉彩顏料上色。這次使用茶色系粉彩，在眼睛邊緣、鼻子和嘴巴的輪廓線上色。

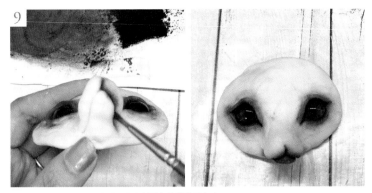

鼻子和嘴巴都上色後，黏貼毛皮布料的事前準備即完成。

從額頭到鼻子都塗上速乾型多用途接著劑。

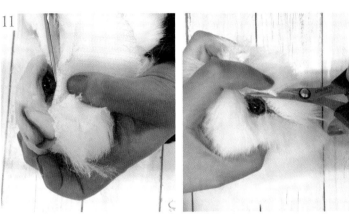

黏貼上毛皮布料。毛皮布料剪下大概的大小後黏貼在頭部，等接著劑乾了之後用剪刀剪去多餘的毛皮布料。

修整多餘的毛。

下巴也貼上毛皮布料後，修整多餘的毛。

眼睛邊緣等細節處修剪毛髮後，再黏貼上成束的毛。

修整完成的樣子。至此上色的事前準備即完成。

16

接著用粉彩顏料為臉部上色。

17
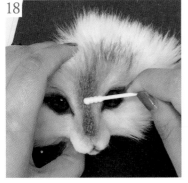

眼線、鼻子和嘴巴的周圍、鼻梁、眼睛上方都
塗上顏色的樣子。

18
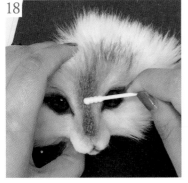

毛流較長的整張臉用棉花棒或牙刷塗上布用塗料。

19

20

接著為鼻子上色。用茶色樹脂黏土（棕色）製作鼻子，並且用速乾型多用途接著劑黏合在臉上。

使用戳針做出鼻孔。

21

等黏土完全乾燥後，塗上水
性壓克力保護漆。

22

準備塞入頭部的
黏土。比對頭部
大小將輕量黏土
搓揉成球狀並使
其乾燥。

115

製作出覆蓋後腦勺的紙型，依照形狀（橢圓形）從毛皮布料裁切下來。因為想呈現圓弧狀，所以在2處剪出牙口後縫合。

用速乾型多用途接著劑，將步驟22的輕量黏土黏合在樹脂黏土。

在輕量黏土塗上速乾型多用途接著劑，黏貼上步驟23的毛皮布料。

從側面看呈現這個形狀。

 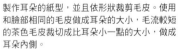

製作耳朵的紙型，並且依形狀裁剪毛皮。使用和臉部相同的毛皮做成耳朵的大小，毛流較短的茶色毛皮裁切成比耳朵小一點的大小，做成耳朵內側。

縫合步驟27的毛皮布料。

準備放入耳朵內的細金屬線。約剪下20cm並且對折，再沿著耳朵輪廓彎折成型。

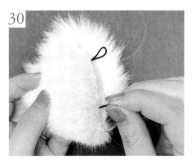

將金屬線放入耳朵後縫合。

31

32

將耳朵縫合在頭部。

一邊耳朵縫合後再縫合上另一隻耳朵,而且需左右對稱、比例適中。至此臉部完成。

····································································

## 身體製作

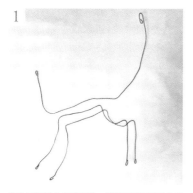

1

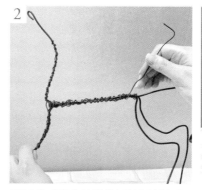

2

用金屬線製作身體骨架。從脖子到尾巴的身體使用一條金屬線,腳左右各 1 條,總共準備 3 條金屬線,並且先將末端彎折捲起。

用比步驟 1 還細的金屬線將身體和腳纏繞組合。接著再捲繞整個骨架加強結構。

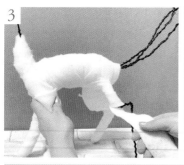

3

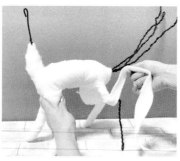

用鋪棉包裹整個骨架塑造整個身形體態。想加粗的部分多裹上幾層鋪棉,增加分量。

Pick up!

日本化線
頑固自在

這是一種材質堅韌又容易摺出形狀的彩色金屬線。形狀不易變形,所以適合用於玩偶的骨架製作。從 1.0mm 到 4.0mm 共有 6 種粗細,所以可以依照玩偶的部位區分使用。

**4**

對照身體和腳的形狀製作紙型。紙型上要標記
毛流方向。

**5**

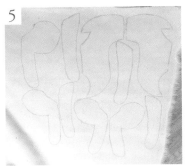

使用和頭部相同的毛皮布料,描繪出紙型後裁切下來。

**6**

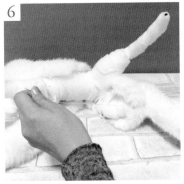

將布料包裹住步驟3製作的骨架後縫合。

**7**

將頭部和身體接合。用剪刀在塞入頭部的黏土
剪出開孔,塗上速乾型多用途接著劑後,將身
體脖子部分的金屬線插入。

**8**

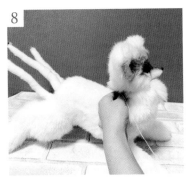

將頭部和身體的毛皮布料縫合。

**9**

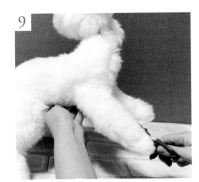

修剪腳部的毛。

**10**

毛修剪完成後,用噴筆從頭部到身體噴塗上布
用塗料。先噴塗當成底色的黃色調。

**11**

接著再噴塗上藍色,身體上色
即完成。

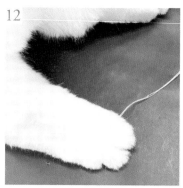

**12**

製作腳趾。從腳尖將縫線刺入並且穿縫 3 次做出內凹的樣子。

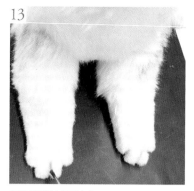

**13**

用粉彩顏料在趾間上色。

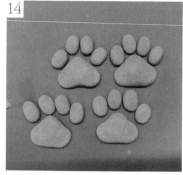

**14**

用樹脂黏土（在MODENA白色黏土加入粉紅色顏料）製作肉球。黏土乾了之後塗上水性壓克力保護漆。

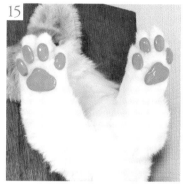

**15**

保護漆乾了之後，在腳底塗上速乾型多用途接著劑。

**16**

尾巴配合身體的底色，使用黃色系的毛皮布料。準備紙型剪出 6 片布料。

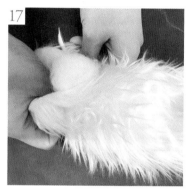

**17**

縫合尾巴並且塞入手工藝棉花。

**18**

將尾巴縫合在臀部。

**19**

用噴筆在尾巴末端和中心線噴上茶色。

**20**

用速乾型多用途接著劑黏上布偶用的鬍鬚部件。在粉彩顏料上色的地方噴上粉彩定畫液噴霧。

**21**

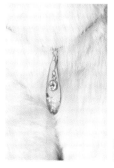

戴上相同設計的耳環和項鍊即完成。

# 龍獸的裝備配件 moruhimeya

龍獸飛行帽／金屬護目鏡／精緻尾部裝飾
覺醒圍巾／堅固包包（附護身符）
附探測功能的雙筒望遠鏡／假日行李箱

創作者運用黏土和人造毛皮等創作出奇妙生物和相關裝備配件。
希望天天都可以創作出似乎具有文明，而且會說話的龍獸和獸人。
作品會不定期在Yahoo!拍賣販售。

[BLOG] http://molsanroom.blog.jp
[Instagram] moruhimeya
[Twitter] @moluxy

# 龍獸飛行帽的製作方法

## 工具和材料

| | | |
|---|---|---|
| ①剪刀鉗 | ⑥粉土筆 | ⑪螃蟹扣 |
| ②手縫線 | ⑦羊毛氈 | ⑫圓環 |
| ③縫針 | ⑧金屬線 | ⑬裝飾部件 |
| ④珠針 | ⑨銅線 | |
| ⑤牙籤 | ⑩鍊子部件 | |

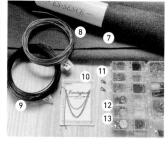

## 製作步驟

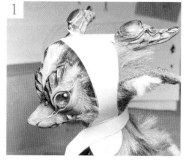

1

在龍頭覆蓋上影印紙,一邊剪貼,一邊製作出紙型。

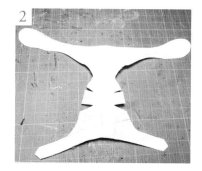

2

為了將包裹頭部等影印紙彎曲的部分攤平,在紙型剪出一些切口(打摺處)。

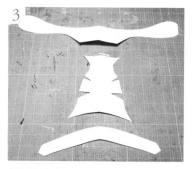

3

紙型裁切完成。

4

也做出耳罩部分的紙型。

5

在羊毛氈描繪出紙型。

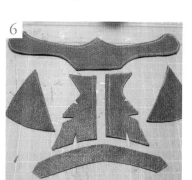

6

預留縫份,在輪廓線外側5mm處裁切。

7

將裁切好的布料正面相對,用迴針縫縫合。

8

燙開縫份,用熨斗按壓燙平。

9

沿著頭部弧線部分縫合打摺處，做出立體造型。

10

將布料正面相對縫合帽簷。

11

下巴扣繩部分用 2 片布料部件夾住本體縫合。

12

耳罩周圍縫上金屬線。金屬線末端較危險，所以先彎折捲起。

13
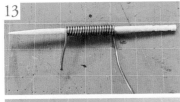

將銅線捲續在牙籤上，如照片所示製作 2 個部件後，縫合在帽子本體。

14

然後穿過下巴扣繩。

15
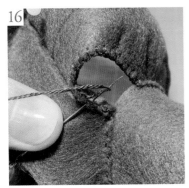

將耳罩縫合。

16
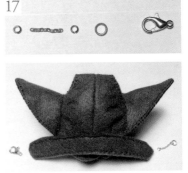

龍角穿出的開孔周圍縫上扣眼繡修飾。

17

將鍊子部件和圓環組合後，縫合在下巴扣繩的左右兩端。

**18**

在下巴扣繩兩端的螃蟹扣和鍊子部件縫上圓珠。

**19**

在下巴扣繩和後腦勺的中央線添加刺繡裝飾。

**20**

在前面縫上裝飾部件裝飾。

**21**

用紅色羊毛氈製作裡布縫合。

**22**

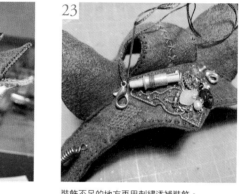

先覆蓋在龍獸上，確定整體比例。

**23**

裝飾不足的地方再用刺繡添補裝飾。

**24**

用金色粗金屬線做成環狀，穿過綠色羊毛氈。

**25**

縫合在飛行帽垂墜處即完成。

# 金屬護目鏡的製作方法

## 工具和材料

①樹脂液　⑥UV燈
②剪刀鉗　⑦羊毛氈
③多用途接著劑　⑧金屬線
④縫針　⑨扁平金屬線
⑤手縫線　⑩裝飾部件

## 製作步驟

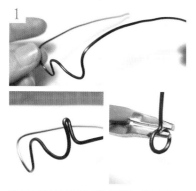

**1** 將金屬線彎折成眼鏡下緣的形狀。彎折兩側，並且將兩側末端彎折捲起以免危險。

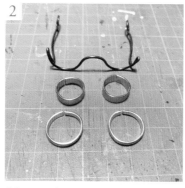

**2** 將寬3mm和寬5mm的扁平金屬線彎成環狀。

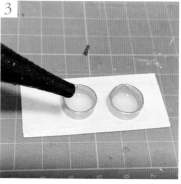

**3** 在寬5mm的環狀內倒入樹脂液，並且用UV燈照射硬化。這個即是護目鏡的鏡片。

**4** 在步驟1彎折的金屬線下側，縫上羊毛氈和步驟2製作的寬3mm環狀。將步驟3的鏡片用接著劑黏在寬3mm的環狀。

**5** 從眼鏡鏡架末端捲繞金屬線，橫越兩個鏡片的上側彎折金屬線。

**6** 將茶色羊毛氈縫合在鼻梁位置。

**7** 在鼻梁部分縫上裝飾部件即完成。

如右圖所示安裝在飛行帽上。

# 精緻尾部裝飾與覺醒圍巾的製作方法

## 工具和材料

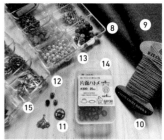

① 縫針　　　⑥ 皮革打洞器　⑪ 齒輪部件
② 手縫線　　⑦ 錐針　　　　⑫ 圓珠
③ 裁縫剪刀　⑧ 皮革　　　　⑬ 木質圓珠
④ 菱斬　　　⑨ 羊毛氈　　　⑭ 扣眼
⑤ 扣眼打孔鉗⑩ 蠟繩　　　　⑮ 裝飾部件

## 製作步驟

## 尾部裝飾製作

1

依照玩偶尾巴的長寬，準備 2 片長方形皮革。
確認添加扣眼的位置。

2

在穿過扣眼的位置開孔，用扣眼打孔鉗夾住，安裝上扣眼。

3

用菱斬開出縫線針孔。

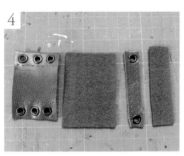

4

縫合裡布。準備比長方形皮革大一些的紅色羊毛氈。

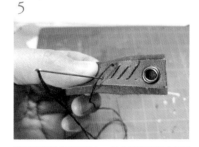

5

將羊毛氈縫合在皮革上，並且將縫線縫成交叉圖案。

6

添加圓珠和齒輪部件。

7

羊毛氈上開孔後穿過蠟繩，並且在末端添加裝飾部件。

**8**

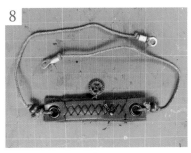

至此條狀皮革的尾部裝飾即完成。

**9**

同樣將羊毛氈縫合在皮革，並且在羊毛氈上開孔。

**10**

穿過蠟繩後交錯編織，並且在末端添加裝飾部件。

**11**

添加裝飾部件後即完成。

# 圍巾製作

## 工具和材料

①莉莉安編織器
②毛線
　（較細的線）
③裝飾部件

**1**

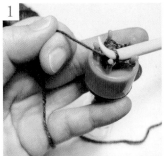

使用莉莉安編織器，以莉莉安編法編織成細圍巾。

**2**

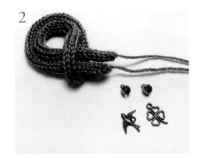

在末端添加鈴鐺、鳥、幸運草的部件即完成。

# 堅固包包（附護身符）的製作方法

## 工具和材料

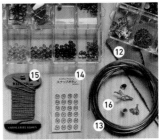

①布用接著劑　　⑦裁縫剪刀　　⑬金屬線
②木工用接著劑　⑧粉土筆　　　⑭按扣
③牙籤　　　　　⑨鉤針　　　　⑮蠟繩
④珠針　　　　　⑩羊毛氈　　　⑯裝飾部件
⑤縫針　　　　　⑪剪刀鉗
⑥手縫線　　　　⑫布料

## 製作步驟

1

配合玩偶的尺寸用圖畫紙製作包包，裁切後製作成紙型。

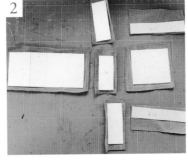

2

在布料上描繪紙型，預留縫份後剪裁布料。

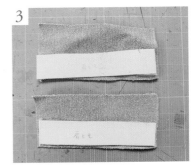

3

肩帶取2倍的寬度。

4

將肩帶對摺後，用布用接著劑固定。

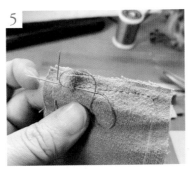

5

將步驟2裁切的布料部件正面相對縫合。

6

開口部分和掀蓋邊緣雙摺邊後，用布用接著劑固定。

7

將縫合的布料翻回正面縫上按扣。

8

為了做出圓鼓狀,將側面捏起縫線。

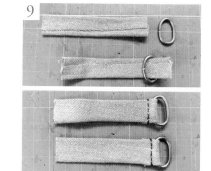

9

用較粗的金屬線做成 2 個環狀,穿過肩帶後縫線固定。

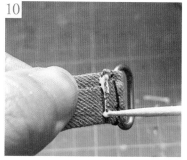

10

邊緣塗上薄薄的木工用接著劑以便防綻。

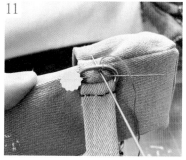

11

將肩帶扣環縫合在包包本體。

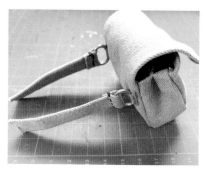

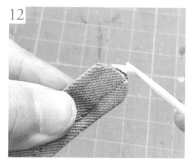

12

肩帶邊緣剪成圓弧狀,塗上薄薄的木工用接著劑以便防綻。

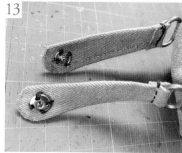

13

加上按扣。

14

將蠟繩編織成掀蓋的提把。

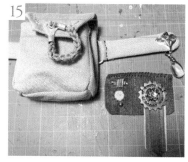

15

肩帶按扣的內側添加水滴狀裝飾。用茶色羊毛氈做成前面的口袋。用裝飾部件裝飾後,縫合在包包本體。

16

在肩帶添加刺繡縫線,掀蓋部分也縫上裝飾部件後即完成。

# 附探測功能的雙筒望遠鏡的製作方法

## 工具和材料

①水性壓克力保護漆
②剪刀鉗
③多用途接著劑
④木工用接著劑
⑤美工刀
⑥雙面膠
⑦縫針
⑧手縫線
⑨繩帶
⑩紙吸管
⑪金屬筒狀部件
⑫環狀部件
⑬樹脂液
⑭UV燈
⑮銅線、金屬線
⑯扁平金屬線
⑰皮革
⑱螺絲、螺帽
⑲壓克力顏料（黑色）
⑳水彩顏料（金色）
㉑裝飾部件

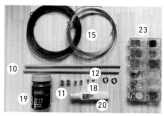

## 製作步驟

### 1

準備2根裁切成3 cm長的紙吸管。

### 2

準備2種粗度不同的金屬筒狀部件各2個，和2個金色環狀部件。

### 3

製作雙筒望遠鏡的鏡片。在環狀部件塗上樹脂液後用UV燈照射硬化。硬化後翻至反面，另一側也同樣塗上樹脂液，並且使其硬化。

### 4

將步驟3製作的鏡片用多用途接著劑黏合在較粗的筒狀部件。

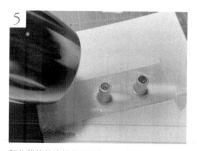

### 5

製作雙筒望遠鏡的眼罩部分。將樹脂液塗在細長的筒狀部件後，用UV燈照射使其硬化。硬化後翻至反面，另一側也同樣塗上樹脂液，並且使其硬化。

### 6

在紙吸管塗上黑色壓克力顏料。

### 7

黑色顏料乾了之後，用水彩顏料（金色）塗出金屬色調。

### 8

用多用途接著劑將步驟4的鏡片和步驟7的筒狀部件黏合。

9

在筒狀部件塗上水性壓克力保護漆。

10

將 2 條寬 5 mm的金色扁平金屬線捲繞在筒狀部件。

11
用寬 3 mm銀色扁平金屬線和螺絲做成裝飾部件。

12
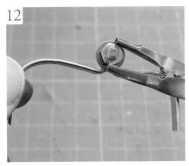
將螺絲和螺帽接合在銀色扁平金屬線突起部分。

13
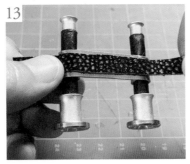
將裁成寬 7 mm的皮革貼在步驟10捲起的金色扁平金屬線上。

14

安裝上步驟12製作的裝飾部件。

15

用銅線固定裝飾的扁平金屬線部分。

16

用裝飾部件裝飾。

17

用手縫線縫上繩帶，以便將雙筒望遠鏡掛在玩偶脖子。

18

為了防綻在繩帶末端塗上木工用接著劑固定即完成。

# 假日行李箱的製作方法

## 工具和材料

① 美工刀
② 尺
③ 木工用接著劑
④ 多用途接著劑
⑤ 牙籤
⑥ 鑷子
⑦ 砂紙
⑧ 雙面膠
⑨ 水性保護漆
⑩ 刷子
⑪ 手鑽
⑫ 鑽頭刃
⑬ 縫針
⑭ 手縫線
⑮ 木板
　（輕木板）
⑯ 鉸鏈
⑰ 金屬扣
⑱ 螺栓、螺帽
⑲ 金屬線
⑳ T 針
㉑ 筒狀金屬部件
㉒ Y 字形金屬
　部件
㉓ 齒輪部件
㉔ 塑膠管
㉕ 羊毛氈
㉖ 裝飾部件
㉗ 皮革
㉘ 羊眼釘
㉙ 轉印貼紙
㉚ 螃蟹扣

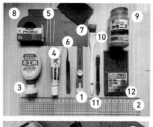
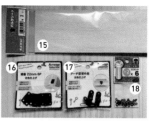

## 製作步驟

1

從輕木板裁切行李箱底部（5cm×8cm）和側面（1.5cm×4.8cm、1.5cm×7.8cm各2片）的板材。輕木板比較軟可以用美工刀裁切。

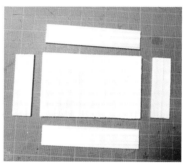

2

側面塗上木工用接著劑後，將每一片側面板材相互組合。

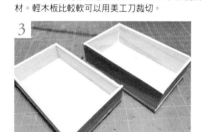

3

以相同方法裁切出行李箱底部（5cm×8cm）和側面（2cm×4.8cm、2cm×7.8cm各2片）的板材，並且用木工用接著劑黏合。

4

用美工刀裁去多餘的部分。

5

整體用砂紙打磨。

6

整體塗上水性保護漆後，使其完全乾燥。

7

將步驟2的箱子在上，步驟3的箱子在下重疊，並且用多用途接著劑黏上鉸鏈。

8

將箱子翻至另一邊的側面，添加金屬扣。

9

用手鑽開孔。因為板材較薄容易裂開，所以要領是使用較細的鑽頭刃小心作業。

10

如果使用金屬扣附的螺絲會穿透內側，所以用較短的螺柱和螺帽固定。

11

行李箱的部分完成。

12

製作滾輪。用塑膠管連接筒狀金屬部件。

13

將金屬線折成三折後扭轉，再放入步驟12的筒狀部件中。

14

在金屬線兩端安裝齒輪部件。這樣就做出可以轉動的滾輪。

15

用多用途接著劑將筒狀金屬部件黏接在行李箱下方做成立柱。

16
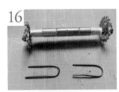
因為要用彎成U字形的金屬線固定步驟14的滾輪部件，所以要在本體的4處開孔。

17

在滾輪部件和U字形金屬線接觸的部分，用雙面膠黏上切成小塊的綠色羊毛氈。

18

將金屬線穿過步驟16的開孔固定滾輪部件。將穿至內側的金屬線彎折捲起。

19
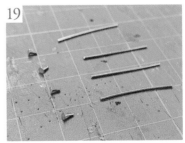
將T針剪短做成小釘子。

139

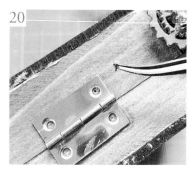

將 T 針做成的釘子，刺進步驟 7 黏接在箱子的鉸鏈開孔。

將螺柱和金屬部件黏接在步驟15的立柱。

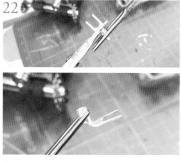

製作行李箱的把手。彎折丫字形金屬部件。

將彎曲的丫字形金屬部件穿過較粗的金屬線。

將整體捲繞上較細的金屬線。

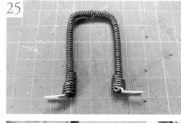

整體都捲繞上金屬線後，將綠色羊毛氈縫在把手部分。

行李箱的基底完成後，接著開始裝飾。

依照行李箱邊角的形狀裁切皮革，並且用接著劑黏貼在四個角。

在四個角添加鉚釘裝飾。

製作可從行李箱的正面拆裝的裝飾。在綠色羊毛氈上添加齒輪部件和裝飾部件的設計。

30

在裝飾部件添加螃蟹扣，在行李箱添加羊眼釘，即完成可拆裝的裝飾部件。

31

用轉印貼紙裝飾。

32

在行李箱內側製作一個盒子，以便收納「附探測功能的雙筒望遠鏡」。用影印紙製作紙型。

33

將紙型描繪在茶色羊毛氈後裁切下來。

34

將側面縫合，做成箱型。

35

剛好可以收納「附探測功能的雙筒望遠鏡」。用緞帶打結固定。製作可以放入其他間隙的小配件即完成。

# 好書推薦

## 飛機模型製作教科書
田宮1/48傑作機系列的世界
「往復式引擎飛機篇」

作者：HOBBY JAPAN
ISBN：978-957-9559-16-4

## H.M.S.幻想模型世界

作者：玄光社
ISBN：978-626-7062-51-7

## 噴筆大攻略

作者：大日本繪畫
ISBN：978-626-7062-08-1

## 骨骼SKELETONS
令人驚奇的造型與功能

作者：安德魯.科克
ISBN：978-986-6399-94-7

## SCULPTORS 06
原創造形&原型作品集

作者：玄光社
ISBN：978-626-7062-31-9

## 依卡路里分級的
女性人物模型塗裝技法

作者：國谷忠伸
ISBN：978-626-7062-04-3

## 大山龍作品集&
造形雕塑技法書

作者：大山龍
ISBN：978-986-6399-64-0

## 飛廉
岡田惠太造形作品集&
製作過程技法書

作者：岡田惠太
ISBN：978-957-9559-32-4

## 動物雕塑解剖學

作者：片桐裕司
ISBN：978-986-6399-70-1

作者介紹

# 綺想造形蒐集室

蒐羅散落於世界各地的奇幻造型作品，企劃活動記錄這些作品的創造過程。著作有『奇幻面具的製作方法（繁中版 北星圖書事業股份有限公司出版）』

## Staff

攝影
田中館裕介
村川寫真事務所

設計、DTP
戶田智也（VolumZone）
株式會社Graphic

企劃、編輯
株式會社Manubooks

編輯統籌
川上聖子（Hobby Japan）

# 幻獸の製作方法
## 可動式幻獸玩偶の製作技法＆作品集

| | | |
|---|---|---|
| 作　　者 | 綺想造形蒐集室 |
| 翻　　譯 | 黃姿頤 |
| 發　　行 | 陳偉祥 |
| 出　　版 | 北星圖書事業股份有限公司 |
| 地　　址 | 234 新北市永和區中正路 462 號 B1 |
| 電　　話 | 886-2-29229000 |
| 傳　　真 | 886-2-29229041 |
| 網　　址 | www.nsbooks.com.tw |
| E–MAIL | nsbook@nsbooks.com.tw |
| 劃撥帳戶 | 北星文化事業有限公司 |
| 劃撥帳號 | 50042987 |
| 製版印刷 | 皇甫彩藝印刷股份有限公司 |
| 出 版 日 | 2023 年 6 月 |
| I S B N | 978-626-7062-53-1 |
| 定　　價 | 450 元 |

國家圖書館出版品預行編目資料

幻獸の製作方法：可動式幻獸玩偶の製作技法＆作品集／綺想造形蒐集室作；楊哲群翻譯. -- 新北市：北星圖書事業股份有限公司，2023.06
　　144 面；　19.0×25.7 公分
ISBN 978-626-7062-53-1（平裝）

1. CST：玩具　2. CST：手工藝

479.8　　　　　　　　　　　111021887

如有缺頁或裝訂錯誤，請寄回更換。

官方網站

臉書粉絲專頁

LINE 官方帳號